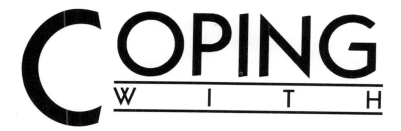

Sexism in
the Military

Mary V. Stremlow, Col. USMCR (Ret.)

THE ROSEN PUBLISHING GROUP, INC./NEW YORK

Published in 1990 by The Rosen Publishing Group, Inc.
29 East 21st Street, New York, NY 10010

Copyright 1990 by Mary V. Stremlow, Col. USMCR (Ret.)

First Edition

Manufactured in the United States of America

Library of Congress Cataloging-In-Publication Data

Stremlow, Mary V.
    Coping with sexism in the military/M.V. Stremlow. -- 1st ed.
        p.   cm.
    Includes bibliographical references and index.
    Summary: Discusses sexism in the military and how to deal with
it.
    ISBN 0-8239-1025-3
    1. United States--Armed Forces--Women. 2. Sexism--United States.
[1. United States--Armed Forces--Women. 2. Sexism.]
I. Title.
UB418.W65S77   1990
355'.0082--dc20   90-37590
                                                      CIP
                                                      AC

ABOUT THE AUTHOR ◇

Colonel Mary V. Stremlow, now a retired Marine reservist, has a bachelor of science degree from New York University at Buffalo. She counts three other women Marines in her family—two aunts who served in World War II and a sister who is a retired major.

Colonel Stremlow served on active duty as a commissary officer, platoon commander, leadership instructor at Officer Candidate School, officer selection officer, inspector-instructor, recruit company commander, battalion S-3, and battalion executive officer.

At the Marine Corps Museum in Washington, DC, she wrote *The History of the Women Marines 1946–1977*. Colonel Stremlow commanded Buffalo's Marine Corps Reserve Mobilization Station at the time of her retirement from the Reserve in 1985. Upon retirement, she was awarded the Legion of Merit.

In 1986 Colonel Stremlow was appointed Deputy Director, New York State Division of Veterans' Affairs. Her responsibilities include planning, supervising, and coordinating the state's veterans programs in twenty-nine counties and assisting the Director in development of Division policies.

Her varied civilian career includes teaching in the Buffalo public schools and employment as a foreman for a major steel company and as an administrator for a defense company.

She has served on the Boards of Directors of the Retired

Officers Association, the Marine Corps Reserve Officers Association, the Marine Corps Historical Foundation, and the Buffalo and Erie County Naval Park. She is Vice-Chairman of the Western New York Committee for Employer Support of the Guard and Reserve and a member of the Department of Veterans' Affairs Advisory Committee on Women Veterans.

Contents

Foreword

I commenced to realize the meaning of sexism in the armed forces while I was a Marine Corps observer with the British army during the Battle for Britain. During a night bombing raid against London, I watched the women gunners in an antiaircraft battery battle the incoming German planes. I suddenly asked myself, "Why aren't our women—able, loyal, and patriotic as they are—permitted to participate in this fashion?"

The answer, of course, was that the sexism that existed in US armed forces at that time reflected the attitude of the majority of the American people. Since then, that attitude has changed for the better, just as has the position of American women in business and politics.

I am convinced that sexism is on the wane in our armed forces. I believe that the time is fast approaching when it will disappear, and all opportunities and billets now available to men will become available to women. It is therefore time to look ahead—to be prepared to seize these opportunities and to fill these billets—if you want a military career.

The author, Mary Stremlow, who as a colonel in the US Marine Corps Reserve wrote *The History of the Women Marines 1946–1977*, continues to sharpen her perspective as Deputy Director, New York State Division of Veterans'

Affairs. She has written here another splendid book that I recommend to you.

Wallace M. Greene, Jr.
General, U.S. Marine Corps (Ret.)
23rd Commandant

Introduction

Is there sexism in the armed forces? Certainly! Can a woman expect to find a rewarding career in the armed forces—one filled with challenge, opportunity, respect, and pride? Absolutely! Sounds a little like double-talk, but I assure you there is no contradiction.

You and I are going to spend some time examining military service for American women, and we are going to concentrate on the matter of gender.

Can you really "... be all that you can be," as the Army recruiting jingle promises? What should you rightfully expect? What will you really find? Will sexist attitudes or policies get in your way to advancement? Just what are sexist attitudes? Will you be a second-class citizen? How do military men feel about serving with women? How about sexual harassment? How will your career be affected by our nation's combat exclusion laws? How will your career be affected by our country's social values? Can servicewomen marry and raise a family?

Before we can answer those questions, we have to agree, at least generally, on what sexism is.

Sexism is present when some persons feel superior to others because of gender. Usually it is associated with the feeling that men are superior to women, although there are instances of sexism in which men are the victims. For example, male nurses have reported career problems resulting from the sexist attitudes of female nurses, doctors, and patients.

In our western culture, most men have been raised to believe that they are different from women because they are larger and stronger. They are therefore viewed as independent, brave, protective, active, and rational. On the other hand, women are usually smaller and more delicate. Thus, they are seen as dependent, fearful, needing protection, passive, and irrational.

Until well into the 1950s "ideal men" and "ideal women" accepted these assigned roles, and those who did not fit into the standard mold were considered odd. Fortunately for all of us, there is much more room for individuality today. We no longer have to be cookie-cutter copies of each other to be accepted and appreciated.

These gender beliefs have influenced the workplace, both civilian and military. If men are the protectors as society believed, it follows that they need greater job security, better promotion opportunities, more money, and more employee benefits. If they are more rational, they are better suited for stressful jobs—such as warfare. The result of this sexist thinking has been fewer opportunities and less pay for women. If you think that these old-fashioned views have been corrected, consider this: Women in 1988 earned only seventy cents for every dollar of a man's paycheck, regardless of skills or education.

As more women went to work, they discovered that they not only faced career obstacles, but they had to fit into a social and cultural environment designed by men for men. The vocabulary of the office or factory is enriched with sports and military jargon. Sophisticated executives can make a business project sound like a battle plan when they talk animatedly about "the troops pulling together" or the branch offices "out in the boondocks." One of the fringe benefits that former military women take with them into civilian life is the ability to speak the accepted language.

Specialist 4 Theresa Pharms of the 101st Division negotiates the "Green Hell" obstacle course at Fort Sherman, Panama. Specialist Pharms was the first female to successfully complete the demanding course (US Army photo released by the Department of Defense).

Business is often conducted at private clubs that do not accept women as members. Many women have been left out professionally because they do not play golf, or if they do, they cannot join in the easy camaraderie of the locker room. It is good to remember that exclusion can be a type of sexism.

Sexism manifests itself in many ways. Sometimes it is obvious, out in the open, bold, and intentional. The signals are very clear and easy to identify: roadblocks to professional success, sexual harassment, derogatory remarks,

lack of respect, less pay, fewer benefits. This type of behavior is the simplest to confront and deal with.

But at other times sexism is hidden, difficult to detect,
perhaps even unintentional. The victim may not even
realize what is happening. The signals are more subtle: an
impressive title without appropriate authority, exclusion
from the decision-making process, not being invited to
lunch "with the boys," never being asked for opinions,
being ignored at business meetings, being excluded from
advanced training.

Without realizing the consequences, a supervisor may
assume that a woman would not want a dirty job and never
offer it to her; a married woman may not be offered a trans-
fer to another city to take a better position; a woman may
not be assigned to a branch office in a crime-ridden neigh-
borhood at higher pay. These sexist assumptions hold
women back without giving them the chance to decide for
themselves.

It would be easy to guess that the armed forces, because
of their masculine image, would be blatantly guilty of all of
the above notions of sexism. On the contrary, they have
done a creditable job of identifying the problems and of
aggressively trying to improve conditions for female members. Many examples of sexism found in the civilian community are no longer an issue of concern for servicewomen.

Pay and benefits are the same for men and women. Pay
is based on rank and length of service. No military woman
will ever learn that a less deserving coworker received a
larger raise in salary because "he is a man with a family to
support."

Responsibility and authority come with increased rank.
Everyone is expected to perform at an acceptable level for
the assigned job. Sergeants supervise corporals, corporals

supervise privates. Gender is never an issue. That was not always the case.

Professional civilian women often are advised to dress in dark, conservative suits to broadcast their position of authority and to gain respect. In the armed forces men and women wear the symbol of their rank on their uniform. Female non-commissioned officers and officers are never mistaken for the girl who makes the coffee.

The military forces have not yet solved all the problems of sexism brought on by our cultural heritage. They have come a long way, but they have a long way to go. Military women are speaking out whenever they perceive inequality, and Congress and the Department of Defense (DOD) are listening. They learned long ago that women are essential to the defense of the United States. There are not enough men to fill the need, and the government estimates that by the year 2000 there will be a severe shortage of enlistment-age men.

Evidence indicates that military leaders are beginning to agree with the 23rd Commandant of the Marine Corps, General Wallace M. Greene, Jr., who wrote in a letter on January 9, 1987, "I have been retired now for nineteen years, but I have never forgotten the part played in wartime and peace by women in the Marine Corps. And I believe that, as time marches on, more and more opportunities and billets in our organization will become available to women. I have always asserted that every activity and specialty in the Marine Corps should be open to women on the same basis as to men."

CHAPTER ◇ 1

History: American Revolution Through World War II

I n 1982, Major General Jeanne Holm of the United States Air Force wrote the book *Women in the Military—An Unfinished Revolution*. There has indeed been a revolution, and as the title so clearly states, it isn't over yet. A young woman who enlists in the armed forces today naturally assumes that she will be trained, employed, paid, and promoted in a way that earlier military women never dreamed possible. And it is safe to say that today's woman can scarcely guess what new opportunities will surface in the next two, five, ten, or twenty years.

History books are filled with stories of women who have fought with armies and on warships. Sometimes they were disguised as men, but not always. It was not unusual for women to be hired as nurses, laundresses, and cooks and to work on the battlefield, enduring all the danger and discomfort of war except direct combat.

During the American Revolution women and girls were part of the militia, and in the Civil War they acted as scouts, spies, and saboteurs. American nurses served in the Revolutionary, Civil, and Spanish-American Wars. Yet there was never a suggestion that women should be considered part of the military.

By the turn of the century the senior officers of the Army were convinced that the nurses were essential to the medical care of our soldiers. In 1901 the Army Nurse Corps was organized, and in 1908 the Navy followed with the creation of the Navy Nurse Corps. These early women were in a semimilitary status. They were neither commissioned officers, nor enlisted. They did not wear rank insignia and did not receive the pay and privileges of their position.

From that time until 1972, when the United States stopped drafting eligible males into the military, advancement for servicewomen was slow or "evolutionary." It always took a crisis, usually a war, to speed up the process. At each war's end, the women returned home. In the 1970s, however, when Congress decided to rely on a volunteer army, it became clear to nearly everyone that the idea could not succeed without womanpower.

At the same time we Americans began to change our views on what the role of women should be. If the feminist movement hadn't come along just as military planners were struggling to maintain a fully manned defense force without the help of the draft, it is possible that military women would still be found mainly in nursing and administrative jobs during peacetime and assigned to the nontraditional jobs only when we found ourselves at war. The pattern was set in World War I.

In 1916, Secretary of the Navy Josephus Daniels correctly assumed that the United States would become involved in World War I, and he knew that he faced a shortage of office clerks. As more and more men would be needed overseas, who would take over the jobs at headquarters and at the many Navy bases? He asked if there was a law that a yeoman (office clerk) must be a man. To the surprise of many, the answer was no.

The Navy Department enrolled 12,500 "yeomanettes," who worked not only in the offices as planned, but as camouflage designers, radio electricians, fingerprint experts, translators, and draftsmen. Some were sent to hospital units in France and to intelligence groups in Puerto Rico.

Later the Marine Corps recruited 300 women known as "marinettes," and the Coast Guard enrolled women to work at headquarters. Only the Army refused to use women as anything besides nurses. The Army Nurse Corps, however, did grow from 400 to 22,000 by the end of the war. Fourteen hundred Navy nurses served at home and abroad.

Military leaders such as the famous General John J. Pershing requested military women to work as clerks overseas, but instead he received civilians and men physically unqualified for combat. The soldiers found that supervising civilians instead of disciplined troops in the war zone was very frustrating.

In all, about 36,000 American women served in the armed forces in World War I. They were adventuresome and bold pioneers who had the courage to leave home and "join up" when many of their sisters still considered it unladylike to work outside the home.

For the first time women received the same veterans benefits as men: government insurance, a $60 bonus, medical treatment for illness incurred in the service, and the privilege to be buried in Arlington National Cemetery.

Only a small number of nurses were kept in the service after the war ended in 1918. All other women were released to return to civilian life. It would be the same for the next fifty years: Bring in women when at war, send them home when at peace. Women were considered temporary fill-ins. Nevertheless, the revolution was under way. The country learned that it could not go to war without its women, and its women wanted to share in the responsibility of defense and freedom.

WORLD WAR II

Primarily because of the efforts of US Representative Edith Nourse Rogers, the War Department in 1941 began to study the possibility of a women's army organization. General George C. Marshall, Chief of Staff, predicted that we would have a severe manpower shortage if the United States entered the war being fought in Europe at the time. He told Congress, "Women certainly must be employed in the overall effort of the nation . . . "

There was much resistance to this radical idea, especially in the House of Representatives, where one member said, "I think it is a reflection upon the courageous manhood of the country to pass a law inviting women to join the armed forces in order to win a battle. Take the women into the armed service, who will do the cooking, the washing, the mending, the humble homey tasks to which every woman has devoted herself? Think of the humiliation! What has become of the manhood of America?"

Eventually, in the absence of other practical alternatives, the military laws were passed and women were soon joining the Army, Navy, Marine Corps, and Coast Guard.

At first women were not considered part of the "men's" army. They were members of the Women's Army Auxiliary Corps (WAAC). It was a bad idea. It didn't work, and it didn't last long. The creation of a separate organization resulted in disciplinary, control, and administrative problems.

Thanks to the interest and intercession of Mrs. Franklin D. Roosevelt, wife of the President, the Navy avoided making the same mistake. On July 30, 1942, a bill was passed in Congress establishing the Navy Women's Reserve. The women were called WAVES: Women Accepted for Voluntary Emergency Service. Women in the Coast Guard were SPARs, named for the Coast Guard motto, "Semper Paratus—Always Ready."

The Marine Corps was the last service to give in to congressional and Navy Department pressure and enlist women. Once the decision was made, however, the Commandant wanted the women to be known as marines and ruled that there would be no nickname for the members of the Marine Corps Women's Reserve. But to everyone they were known as WRs, standing for women reservists.

These titles, WAACs, WAVES, SPARS, and WRs, are a proud part of the military history of American women. They are used even today in informal conversation and as names for women veteran's associations. But some military historians and sociologists believe that the nicknames implied that women were somehow less than soldiers, sailors, marines, and coast guardsmen. That is why, in the spirit of integration of women into the armed forces in the 1970s and '80s, there was a concerted effort to do away with the nicknames.

During World War II, 350,000 women served in the

armed forces. That was approximately 2.3 percent of our military strength. At first most people were skeptical about the types of jobs women could perform. There was heated debate about which military occupational specialties were suitable for women. It was generally believed that most nonclerical jobs could be done satisfactorily only by men.

The War Department conducted a study and proved them wrong. It indicated that one in three army jobs could be assigned to women, and that the army could enlist 1.5 million women without endangering its fighting capability. The jobs that were considered inappropriate for women involved:

- combat
- "unsuitable" working conditions
- long training periods.

Of these notions regarding the use of women in the armed forces, two are still argued today—should women serve in combat? and are women physically strong enough for certain jobs?

By today's standards the study was certainly sexist. For example, all supervisory positions were considered closed to women: It was unthinkable that women should be in charge of men. Personnel jobs involving classification of men for combat duty were declared unsuitable because it was thought that men would be resentful. These two issues are no longer a problem. Women supervisors and commanders are an accepted fact, and the field of personnel classification has long been open to women.

The Marine Corps tackled the task of separating "women's" work from "men's" work a little differently. They created the following four job categories based on their idea of what females could do:

CLASS I: Jobs in which women are more efficient than men. Examples: all clerical jobs, especially those involving typing or fairly routine tasks but requiring a high degree of accuracy; administrative jobs connected with organization and administration of the Women's Reserve; and instructional jobs of all types.

CLASS II: Jobs in which women are as good as men and could replace men on a one-to-one basis. Examples: some clerical jobs in which men are especially good, such as accounting; some relatively unskilled service or clerical jobs, such as messenger or Post Exchange clerk; some of the mechanical and skilled jobs, such as watch repairman, fire control instrument repairman, tailor, sewing machine operator—especially those jobs requiring a high degree of finger dexterity.

CLASS III: Jobs in which women are not as good as men but can be used effectively when the need is great, such as wartime. Examples: most of the jobs in motor transport— men are better as motor mechanics and even as drivers when the equipment is heavy and the job also demands loading and unloading, as it often does; most of the "mechanical" and "skilled" jobs; supervisory and administrative jobs, such as first-sergeant (except in WR units), where maximum proficiency depends on years of experience in the Marine Corps, and also some supervisory jobs where part of the personnel being supervised is male; strenuous and physically tiring jobs, such as mess duty, where experience showed that more women had to be assigned to cover the same amount of work because they could not endure the long hours and physical strain without relief as well as men.

CLASS IV: Jobs in which women cannot or should not be used at all. Examples: jobs demanding excessive physical strength, such as driving extremely heavy equipment, stock handling in warehouses, heavy lifting in mess halls; jobs totally inappropriate, such as battle duty or jobs requiring that personnel be engaged at particularly unfavorable hours; jobs protected by special Civil Service regulations for civilians, such as librarians.

The Navy, on the other hand, considered all jobs open to women, and commanders were instructed to make decisions on an individual basis.

Despite the restrictions imposed by the Army's study and the Marine Corps' classification system, women moved easily into the "unsuitable" jobs. There was no choice. As great numbers of men were sent overseas to fight, women replaced them. They served as truck drivers, mechanics, parachute riggers, postal workers, photographers, and naval air navigators. They were in great demand in aviation units, where the men have always seemed less conservative in their views.

During World War II women accepted the call to "Free a Man to Fight," and they left their mark on American military forces. The idea of "women's work" has never been the same. The problem was not what can they do? but can we recruit all we need? There was even some talk of drafting women.

One reason the women of World War II were so successful was the high recruiting standards set for them. Women had to be older, better educated, more skilled, and of higher moral character than male recruits or draftees. Of course, that was blatant sexism, but the women did not complain. On the contrary, most believed that their acceptance by military men and the American public depended upon superior performance and good character. History proved them right. This is an example where sexism probably helped women to advance.

In later years when the women's enlistment standards were lowered to match the men's so that both sexes would receive equal treatment, the military women screamed the loudest. They believed that they had to be better to get ahead and to maintain whatever foothold they had. Like many working women today, they believed that a woman who didn't measure up to their ideals was a poor reflection on the performance of all women.

At the end of World War II most Americans expected the servicewomen to return home, and the process was begun.

CHAPTER 2

Career Status: 1948

The task of demobilizing the Army, Navy, Marine Corps, and Coast Guard after the war was enormous. Of most significance to women was the realization that this was essentially an administrative process requiring more clerks than combatants. There is an old saying that an army fights on beans and bullets. In 1945 the War Department learned that an army goes home on a mountain of paperwork. Women became more indispensable, not less. Policies regarding the discharge of women changed daily. Even men who were strongly opposed to the idea of women in uniform begged to keep their female clerks to assist in processing the untold number of documents needed to release the 12 million people in the US Armed Forces.

In the Marine Corps Museum is an undated, unsigned, brief history of women in the Marine Corps that begins, "It is rumored that when it was announced that women were going to be enlisted in the Marine Corps the air was colored with profanity in the language of every nation as the members of the old Corps gathered to discuss this earth-shattering calamity. It is entirely probable that the wailing and

moaning which went on that day amongst the old Marines
was never equaled—never, that is, until it was announced
that the women Marines were going home. Then, with a
complete reversal of attitude, many of those same Marines
declared that the women in their offices were essential
military personnel and absolutely could not be spared from
the office."

Disagreement arose among the services and even among
the women wartime leaders about whether women should
stay or leave. If they were to stay, nearly everyone agreed
that their status would have to be more like the men's.
Competent women who needed to work would not remain
in the service without some guarantee of job security,
reasonable benefits, and privileges. In fact, a large number
of skilled women who wanted a military career returned to
civilian life because of the uncertain status accorded them
from 1945 to 1948. They simply could not afford to wait.

To appreciate this time in military women's history, it
is necessary to understand the difference in the terms
"regular" and "reserve." In the armed forces, regulars are
people who intend to stay on active duty at least twenty
years and then retire with a pension. They are the career-
ists. Reserves usually spend several years on active duty
and then return to their civilian occupation. They can
continue their affiliation by joining a reserve or National
Guard unit in their hometown and attending monthly
training sessions. In time of war, reservists and guardsmen
are ordered to active duty.

Until 1948 all women were reservists and therefore
enjoyed none of the privileges or benefits of regulars.
They had no job security, no pension plan, and served
under very restrictive policies regarding assignment and
promotion.

General Dwight D. Eisenhower, who at war's end

became Army Chief of Staff, led the fight to make a place for women in the regular forces. Like many high-ranking officers, he recognized the invaluable contribution made by the women. More than that, he realized that women would always be needed in time of war and that it was grossly inefficient to start up a new organization at the onset of every world crisis.

The Navy followed close behind and announced plans to keep women beyond the original demobilization date. The Marine Corps was a reluctant partner. The Marines wanted to establish reserve units of women throughout the country, keeping the option to call them to active duty only if necessary. The Corps didn't want anything to do with full-time career women. Only the Coast Guard immediately released all its women.

For nearly three years the plan to give women career status floundered in the Congress. Representative Margaret Chase Smith helped to shepherd the legislation through to passage. The arguments against the bill often reflected the common but accepted sexist attitudes of the day. For example, some legislators wanted to make it mandatory that women retire at a younger age than men because they believed women were prone to become unstable when they reached the age of menopause. Others were fearful that the husbands of career servicewomen would be lazy, content to follow their wives from camp to camp and take advantage of the benefits for military dependents. In fact, the final version of the bill included specific provisions to minimize this threat. Similar restrictions were never placed on the dependents of men in the service.

On June 12, 1948, President Harry S. Truman signed the long-debated Women's Armed Services Integration Act, Public Law 625.

Generally, P.L. 625 gave equal status but not equality to women in uniform. It contained a number of restrictions and special provisions. General Holm, in *Women in the Military—An Unfinished Revolution*, summarizes the most significant provisions of the law as follows:

- Gave permanent status to women in all armed forces—Regular and Reserve.
- Gave permanent status to the Women's Army Corps as a separate organization within the Army, "which shall perform such services as may be prescribed by the Secretary of the Army," and required all women not members of the Nurse or Women's Medical Specialist Corps to be members of the WAC. No similar organizational entities were established in the other services.
- Authorized the appointment of female commissioned officers and warrant officers and the enlistment of women in the Regular and Reserve components of the four services.
- Imposed a 2 percent ceiling on the proportion of women on duty in the Regular establishment of each service.
- Limited each service to only one line full colonel or Navy captain. (No generals or admirals were allowed at all.) This senior grade could be held only for a temporary period of four years unless extended by the service Secretary. The Army colonel had to be "the Director of the Women's Army Corps"; the Navy captain was required to be "an assistant to the Chief of Naval Personnel" (presumably for women); and the Marine Corps colonel was assigned to assist the Commandant in the "administration of women's

affairs." Only the Air Force was unconstrained in the assignment of the one authorized full colonel.

- Set a 10 percent limit on female officers who could serve as permanent Regular lieutenant colonels and Navy commanders. In the case of the Navy, a 20 percent limit was imposed on the number of lieutenant commanders.
- Established separate officer promotion lists for women in each grade in the Army, Navy, and Marine Corps. Only in the Air Force were women integrated into the male promotion lists in all grades below colonel.
- Set the minimum enlistment age at eighteen, with parental consent required under twenty-one (as compared to seventeen for men, with parental consent required under eighteen).
- Provided that officers and enlisted women could claim husbands and/or children as dependents only if it could be proved that they were in fact dependent upon women for "their chief support." Wives and children of male members were automatically considered dependents.
- Authorized the service Secretaries to prescribe the military authority that women might exercise and the kind of military duty to which they might be assigned provided, in the case of the Navy and Air Force, that they "may not be assigned to duty in aircraft while such aircraft are engaged in combat missions"; nor, in the case of the Navy, "may they be assigned to duty on vessels of the Navy except hospital ships and naval transports."
- Authorized the service Secretaries to terminate the Regular commission or enlistment of any female

member "under circumstances and in accordance with regulations proscribed by the President." No such blanket authority existed for discharging men.

At the time, P.L. 625 was considered a victory for women. Looking back, it was sexist legislation. It allowed the services to administratively discharge women for unspecified reasons; it permitted women with dependent children to be discharged without exception—even if the children were teenaged stepchildren who resided part time in the servicewoman's home; it denied married women the right to government family housing. These are but a few of the provisions that irked and angered servicewomen for twenty years. Only in the late 1960s did some of the barriers begin to crumble.

Unfortunately, the momentum and enthusiasm of the women dwindled during the long-drawn-out process of passing this flawed legislation. The reaction was disappointing, and the response was poor. To add to the deteriorating situation, the draft was reinstated just weeks after the passage of P.L. 625, thereby making the need for women less critical. The effort to integrate and accommodate their special needs was lukewarm. Once again, women were relegated to the sidelines until the next war.

CHAPTER 3

A Revolution in Reverse: 1950s and 1960s

"The calm before the storm" is one way to characterize women's military programs from 1950 until the mid-sixties. It was a time of dissatisfaction and frustration for serious career-minded women. It was as if all the male military leaders forgot the lessons learned in World War II. Some of the most belittling and patronizing notions, quotations, and studies are to be found during this period of stagnation.

When the war in Korea broke out, approximately 22,000 women were in uniform, of whom 7,000 belonged to the health services. The Department of Defense planned a recruiting campaign to raise that figure to a total of 112,000, but the number peaked at 48,700 in 1952.

Several explanations are given for this dismal failure.

One obvious factor is that the war in Korea never fired the patriotism of the country as did World War II. The fervor just wasn't there. Then, too, Americans were still ambivalent about the idea of women in uniform. If anything, they were more apt to question the femininity and morals of a woman who enlisted.

The excessively stringent recruiting qualifications didn't help. As was the custom by now, women had to be better educated and were subject to higher mental, physical, moral, and emotional standards than men. Unreasonable investigations were conducted to weed out "misfits" who were judged incapable of adjusting to military regimentation and communal living. The Air Force went so far as to require that every female enlistee be given a psychiatric examination.

The most positive outcome of this era was the establishment of the Defense Advisory Committee on Women in the Services (DACOWITS), a group of fifty prominent women who were asked to advise and assist the Secretary of Defense on the matter of recruiting women. From 1951 until the present day DACOWITS has played a leading role in improving the status of women in the military.

While the programs were in great jeopardy, women continued to serve with distinction, often assigned to boring jobs for which they were overqualified. They never comprised more than 1 percent of the total force—half of the number allowed by law.

Nurses were the exception. The need for them was immediate and recognized. They landed in Korea only four days after the first American troops. Many of these women were veterans of World War II who had joined the reserves and were recalled to active duty in 1950.

During the decade after the war in Korea, there was no

progress. General Holm termed it a " . . . period of survival in an unwelcome environment." The number of women on duty declined from 35,000 in 1955 to 30,600 in 1965, despite the increasingly vocal feminist movement and the progressive policies of President John F. Kennedy. The President in 1961 said, "If our nation is to be successful in the critical period ahead, we must rely upon the skills and devotion of all our people. In every time of crisis women have served our country in difficult and hazardous ways. They will do so now, in the home and at work . . . . Women should not be considered a marginal group to be employed periodically only to be denied opportunity to satisfy their needs and aspirations when unemployment rises or a war ends."

The women in uniform were losing ground at a disturbing rate. There was talk of giving up the concept of women in the armed forces. They were squeezed out of one technical field after another. Skill training became a rarity, and then only if it could be accomplished in a short time. Professional development training was nonexistent. Dissatisfaction was rampant, and large numbers of women left the service before the end of their enlistment—usually based on the marriage escape policy.

All this reinforced the view that women were too much trouble and too costly to recruit and train. It led to a Government Accounting Office (GAO) study in 1963, which concluded that the military services must find a way to reduce the turnover rates of women or consider alternative sources of manpower such as Civil Service employees.

The damaging GAO report and another manpower crisis, this time caused by the war in Vietnam, served as a stimulus to take a good, hard look at the issue. All services reviewed their discharge policies and searched for ways to improve the retention of women. A large part of the problem was obviously the excessively liberal discharge policy upon marriage.

The Marine Corps can serve here as an example. From 1949 until 1965 enlisted women Marines could request a discharge when they married as long as they had completed one year of their enlistment beyond training. Officers had to wait two years. This impractical system stemmed from society's negative attitude toward working wives. In 1964 the Director of Women Marines, Colonel Barbara J. Bishop, decided that women must honor their enlistment contract. To make it easier, it was announced that whenever possible military couples would be stationed at the same or nearby bases. Until that time, keeping couples together was not considered important.

One year later the policy of discharge based on marriage was suspended. The desired result was realized almost immediately. The rate of discharges for marriage was dramatically reduced from 18.6 percent in 1964 to 6.3 percent in 1965 and 2.3 percent in 1966.

During this era of reexamination of the concept of women in the armed forces, President Lyndon B. Johnson signed Public Law 90–130, which removed restrictions on the careers of female officers in the Army, Navy, Air Force, and Marine Corps. Most of the attention was given to the fact that women could now be promoted to general and admiral, although no one seriously expected that to happen in the foreseeable future. Little mention was made of the removal of the 2 percent ceiling on the number of women who could serve; again, no one expected the services to exceed the previously prescribed limit. As it turned out, P.L. 90–130 made possible the giant strides achieved in the seventies and eighties after the draft ended and the United States came to rely on an all-volunteer force.

VIETNAM

According to Defense Department figures, approximately 7,500 military women served in Vietnam. A US Senate report states that there were 36 women Marines, 421 women in the Navy, and 771 in the Air Force. The remainder were in the Army. Army, Navy, and Air Force nurses accounted for 80 percent of the total.

Hundreds more could have been used in noncombat positions in Southeast Asia, thereby making more men available to fight. Some argue that if more women had been sent overseas, the armed forces would not have been obliged to lower their standards in order to fill their conscription quotas. But that is where we paid the consequences of the neglect and apathy of military planners during the preceding fifteen years.

Not enough women were available, and they had not been trained in the skills needed. At one point an Army commander in Vietnam requested seventy WACs to work with sophisticated data processing equipment, but only twenty with the necessary qualifications were available.

Even more difficult to understand is the fact that the military had never made plans for sending women to a combat area. They were completely unprepared. The women did not have appropriate or practical uniform items for the work, area, or climate. The Pentagon had no system for identifying positions women could fill overseas. Women had not been given rudimentary training with weapons and could not even defend themselves.

Senior male commanders recognized the need for more military women but were stymied by the bureaucratic mess. They pointed out that nurses, Red Cross workers, and other civilian women were working at nearly every base and especially in the headquarters commands in

Saigon. Why not military women? This caused a morale problem for both men and women. The women were eager to serve: That was why they had enlisted in the first place. It was demeaning to be shunted aside. Men, on the other hand, were resentful. They rightly asked, what good are you if you're not going to be here when we need you?

This is a good example of sexism by exclusion. We can find no policy that specifically stated that women could not or should not serve in Southeast Asia. Women had simply been left out of the planning process. Many people, men and women, believed that women did not have the endurance to survive in a dangerous, dirty environment without modern conveniences. They ignored the fact that nurses in the field served courageously under the most primitive conditions. An Air Force sergeant who had been denied an assignment to Vietnam told the WAF Director, "If American women were half as fragile as the brass seem to think they are, we never would have conquered the West."

Approximately 1,300 women of the line (non-nurses) served in Vietnam. They were secretaries, clerks, air traffic controllers, photographers, cartographers, communication specialists, and information specialists. They performed top secret work in the intelligence field, and some even helped organize and train the Vietnamese Women's Armed Forces Corps.

Navy nurses lived and worked on either the USS *Repose* or the USS *Sanctuary*, hospital ships that sailed off the coast of Vietnam, normally between Danang and the Demilitarized Zone. Air Force flight nurses cared for the wounded evacuated by plane to the United States, Okinawa, Japan, or the Philippines. Army nurses staffed field, surgical, and evacuation hospitals as well as MASH (Mobile Army Surgical Hospital) units from Quang Tri in the north to Can Tho in the south. These women were in combat zones; the hospitals were often under attack; yet because of society's attitude toward women being "the protected," they had no weapons or training in self-defense.

Four Navy nurses were injured during a Viet Cong terrorist attack. Eight Army nurses were killed, mostly in air crashes, but 1st Lt Sharon A. Lane died of shrapnel wounds during an enemy rocket attack on an evacuation hospital. Captain Mary T. Klinker of the Air Force was killed while serving as a flight nurse on a C-5A that crashed on takeoff while evacuating a load of Vietnamese orphans just before the fall of the Saigon government.

The lessons of the war in Vietnam, as they apply to the use of women in the military, are clear and have been recorded by feminist historians. If women are to be allowed to make a real contribution to the defense of the United States, they must be:

- recruited in adequate numbers;
- trained in a broad range of skills;
- given professional development training on a par with the men;
- allowed to fill positions that lead to promotions and greater benefits;
- included in strategic military plans.

CHAPTER 4

An Idea Whose Time Had Come: 1970s and 1980s

The 1970s and 1980s were the breakthrough years; by the end of the seventies the United States was by far the world leader in the use of military women. In terms of real opportunity, the armed forces had outdistanced the private sector and could serve as a model in critical areas such as pay, promotion policies, training, professional development, travel, and benefits.

A combination of events came together at the right time. The end of the draft created an immediate need for added sources of "manpower." Congressional passage of the Equal Rights Amendment (ERA) sent military planners scurrying to resolve problems of sex discrimination. Also, in an age of acute awareness of civil rights, servicewomen went to the courts to challenge the constitutionality of policies on equal protection grounds.

President Richard M. Nixon ended the draft in 1973, but the idea had been studied for several years before. A congressional report on the utilization of military manpower stated:

"We are concerned that the Department of Defense and each of the military services are guilty of 'tokenism' in the recruitment and utilization of women in the Armed Forces. We are convinced that in the atmosphere of a zero draft environment or an all-volunteer military force, women could and should play a more important role. We strongly urge the Secretary of Defense and the service secretaries to develop a program which will permit women to take their rightful place in serving in our Armed Forces."

The services were asked to come up with plans to double their enrollment of women by 1977. The skeptics were appalled, but in fact an unprecedented thing occurred— the services exceeded their own goals, and the number of women rose from 45,000 in 1972 to 110,000 in 1977.

The services experienced difficulty in attracting enough qualified men to meet the demand and found that by recruiting more women they could improve quality and save money. More than 91 percent of the women recruits were high school graduates, compared to less than 67 percent of the men. All of the female recruits were average or above average in intelligence. Further, it was learned that the Army spent $3,700 to recruit a high-quality male versus $150 for a similar female. Even the Air Force, showing the least difference in cost, spent $850 for males compared to $150 for females.

Although all the task forces, studies, and congressional hearings in the seventies came to the same conclusion—

that the armed forces could use many more women and still comply with the combat exclusion laws—the military leadership found a number of justifications to lower the projected figures.

The Navy declared that because of shipboard requirements and the need to have shore jobs available for seagoing sailors to rotate into, they could not enlist women based solely on the number of noncombat skills open to them. If all the shore jobs were held by women, there would be no alternative but to lengthen the sea time of men. The Air Force, which by one estimate could increase its female strength to 76 percent, reasoned that a lack of adequate facilities was a problem. Someone decided that it was not appropriate for single men and women to live in dormitory-style quarters sharing the same hallway.

Tests were conducted to determine if the number of women in a military unit would cause efficiency to go down. There was, and still is, a theory of "male bonding" that casts a doubt over the social impact of women as teammates in a stressful situation. The Army found that properly trained and led women were good soldiers.

An analysis prepared for the Defense Department concluded:

"The tradeoff in today's recruiting market is between a high-quality female and a low-quality male. The average woman available to be recruited is smaller, weighs less, and is physically weaker than the vast majority of male recruits. She is also much brighter, better educated (a high school graduate), scores much higher on the aptitude tests, and is much less likely to become a disciplinary problem.

"To put the question bluntly: Is recruiting a male high school dropout in preference to a smaller, weaker, but higher-quality female erring on the side of national security, in view of the kinds of jobs which must be done in today's military? The answer to that question is central to the decision on how many women should be used in the various services. Sometimes the answer will be yes, and sometimes it will be no, but the question continues to be relevant."

So, once again, the Secretary of Defense ordered that the goals for women be doubled. After the inception of the All-Volunteer Force, the United States witnessed an explosive increase in the number of women in the military services from 55,402 (2.5 percent) in 1973 to nearly 222,000 (10.3 percent) in 1988.

As the numbers increased, so did the opportunities, and that in turn made the military more attractive to women. The 1970s and '80s were decades of rapid and far-reaching policy changes that recast the status of American military women.

Reserve Officers' Training Corps (ROTC) was opened to females, making scholarships and equal training available. By 1979, 20,000 women were enrolled, and it became the primary source of female officers.

Flight training in the Army and Navy was opened in 1973 and in the Air Force in 1976.

By 1975 the Navy had named its first woman flight surgeon, and women served on service craft (e.g., tugboats). Marines were assigned to selected Fleet Marine Force Commands—the combat units of the Corps.

In 1976 women were admitted to the service academies, and the Marines allowed women to participate in field training exercises.

Training was integrated in all services to varying degrees. Weapons familiarization was introduced. The traditional system of separate women's units was abolished, and women were integrated into the male organization.

The nicknames dating back to World War II were dropped, and their use was discouraged. The women were told that they no longer had to use designators such as Women Marines, WAVES, WACs, WAF. They became marines, sailors, and airmen.

Promotion systems were consolidated, and separate promotion lists were abolished in the Army, Navy, and Marine Corps. The Air Force had from the start followed an integrated procedure. That is a mixed blessing in the view of many, because until combat restrictions are lifted entirely, men will continue to accumulate operational experience that skews the process in their favor.

Finally, each of the services eliminated the office of women's director. The women were to be treated, managed, and administered in the same way as men.

DoD rewrote directives to say that women were to be treated equally with men regarding dependents, and the word "wife" was changed to "spouse." The services began to grant waivers to women who wanted to remain on active duty and raise a family. Maternity uniforms were designed, much to the consternation of old-timers, men and women. In 1988 the Army began to issue maternity camouflage field uniforms.

Men and women who served before 1970 would have a hard time recognizing the armed forces today. To be sure, problems such as assignment to combat positions and sexual harassment linger and beg to be resolved. As in the civilian world, the services are struggling with issues of child care, availability of adequate health care, and quality of life. But in the eyes of many the American military has moved to a position of world leadership in the employment of women. It's a career worth exploring.

CHAPTER 5

The Combat
Dilemma

ombat exclusion laws and service policies restrict women from so many career opportunities that most servicewomen would say this is the last important barrier to be challenged. Promotion to the top ranks for both officers and enlisted depends very much on the experience of command. Women are not prohibited from command positions, but they are barred from many of the combat and combat-related fields where the majority of command jobs are found.

The opposition to women's serving in combat stems from society's views on the proper role for women and from the reluctance of military leaders to cross this controversial bridge. Congress made it official in the Armed Forces Integration Act of 1948:

> Section 8549 of 10 USC prohibits the permanent assignment of female members of the Air Force to duty in aircraft engaged in *combat missions*. Excep-

tions are made under Section 8067 for medical, dental, chaplain, and other "professionals."

Section 6015 prohibits the permanent assignment of female members of the Navy to duty on vessels or on aircraft that can be expected to be assigned a *combat mission*.

Section 3012 provides that the Secretary of the Army may assign, detail, and prescribe the duties of the members of the Army. The restrictions imposed upon the Navy and Air Force are conspicuously absent. This is the way the Army wanted it when the law was written in 1948.

The Marine Corps falls under the Department of the Navy and follows the restrictions of Section 6015. In addition, the Marine Corps further restricts women from serving in combat units or "combat situations."

For nearly thirty years the notion persisted that women were outlawed from combat and all things related to combat. Few people bothered to read the law with any real scrutiny until the manpower crisis brought on by the war in Vietnam and later by the end of the draft. Unknowing servicewomen accepted the myth that they could not legally serve in most of the nontraditional fields and at many of the exciting posts and stations around the world.

A careful reading of sections 8549 and 3012 of US Code 10 clearly reveals that the prohibitions were restricted to aircraft and ships engaged in or assigned to combat missions. The services have spent many man-hours trying to define "combat mission." Unfortunately, each branch has come up with its own interpretation and its own set of rules on how women can be used in combat and combat situations.

Captain Sue Gillespie readies the largest aircraft in the Free World—the C-5A Galaxy—for an overseas flight. Gillespie was the first female to become an aircraft commander in the Air National Guard's 105th Military Airlift Group. In 1989 she routinely flew supply missions between Ramstien air base in West Germany and Stewart Air Base in New York.

A Coast Guard petty officer navigates a busy port (official US Coast Guard photo).

In 1988 the Pentagon took a step toward making better sense of a woman's role in the military. A special task force began the process of developing rational, consistent criteria for opening all jobs except those that are strictly combat. The fresh approach was essential because combat is not carried on the way it was when the 1948 law was written. It no longer occurs on clearly defined front lines with soldiers battling face to face and hand to hand. The art of war has gone hi-tech. Many combatants are miles to the rear, not on the front line, but certainly in areas expected to be high-priority targets. Arthur Hadley in his book *The Straw Giant* states that in future wars women will be casualties out of proportion to their numbers because of the critical jobs they will fill in the rear areas. The year-long Pentagon study resulted in a new definition of "combat mission," which was announced to the services along with an order to use it in redefining the use of their women.

Combat mission was defined in 1988 by the Secretary of Defense as: "A task, together with the purpose, which clearly requires an individual unit, naval vessel, or aircraft to individually or collectively seek out, reconnoiter, and engage the enemy with the intent to suppress, neutralize, destroy, or repel that enemy."

As an outcome of the work of the task force, combat positions for women are now based on a degree of "risk," and 24,000 new positions were made available to women in 1988.

The following comparison illustrates some of the policies of assigning women to ships, planes, missile crews, and positions on the battlefield:

Ships

Coast Guard women serve on all ships in the Coast Guard fleet and in all positions, including command of ships. The Coast Guard has no restrictions based solely on gender in assignment, training, recruitment, or career opportunities.

The Coast Guard is part of the Department of Transportation in peacetime but would come under the control of the Secretary of the Navy in time of war. The Coast Guard does not consider itself bound by the statutory restrictions that prohibit the Navy from assigning women to certain ships.

The Navy assigns women to repair and research ships and to those that perform a noncombat mission, such as shuttle ships that deliver supplies from base supply to the battle group. The high-speed support ships that remain with the battle group are closed to women.

The Navy is also supported by the Military Sealift Command (MSC), which consists of civilian ships under contract to the Navy. Civilian women regularly sail on the MSC ships serving with the Navy's battle group. At times military personnel are included on the civilian MSC ships, and in 1986 the Navy opened these permanent assignments to Navy women.

Aircraft

Women can be pilots and navigators in the Army, Navy, and Air Force, but there are differences regarding their assignment to various types of aircraft. In general, women fly combat support aircraft such as cargo planes, refueling planes, AWACs, weather planes, medical evacuation planes, and certain helicopters. However, they are barred from permanent assignment to fighter jets.

In 1986 the Navy designated a woman helicopter test pilot and a year later selected a woman pilot to be commanding officer of a shore-based squadron that specializes in electronic warfare.

Missile Crews

In the Air Force women can be assigned to all missile crews. Before 1985 they were barred from the two-person crews of the Minuteman missile system. The policy change requires that both crew members be of the same gender.

Short-range missile systems, which are usually located close to the field of battle, are closed to Army women. They are assigned to long-range missile systems such as the Pershing and Nike-Hercules, located in the rear.

In the Navy, missile systems are found on submarines, which are closed to women.

Battlefield Location

The Army is unique is its use of the Direct Combat Probability Code, a system that determines the probability of engaging in direct combat for every position in the Army. Women are barred from serving in positions coded P1, but they may be assigned to combat service support positions that may routinely bring them into the P1 location on the battlefield.

All this is not meant to confuse you; neither should it discourage you from seeking a career in the armed forces. What should be obvious is that the law is not nearly as restrictive as many people believed for the past forty years. Too often military and congressional leaders used the law as a needless barrier, thereby keeping women out of nontraditional occupational fields.

As recently as 1986, Secretary of Defense Caspar W. Weinberger explained his opposition to women in combat, saying that it " . . . may be an unpopular stand to take but I think it's proper for you to know about it. Either I'm too old-fashioned or something else is wrong with me, but I simply feel that that is not a proper utilization. And I think, again to be perfectly frank about it and spread all my old-fashioned views before you, I think women are too valuable to be in combat."

The Secretary's statement is supported by oft-repeated myths regarding women in general. Some would argue that a woman's performance, whether in business or combat, is limited by lack of physical strength, the capability to become pregnant, menstruation, and the assumption that women are less able to work under stress. Further, they point out that the presence of women in a combat unit interferes with "male bonding," defined as the tendency of men to draw together and to reject females in situations like war and police work.

It is true that the average man is stronger than the average woman, but not in all cases. Therefore, it is sexist and discriminatory to screen out all women from certain occupational fields because of gender alone.

Pregnancy is seen by some as a severe handicap to working women. Care must be taken not to assume that all pregnant women must be reassigned as soon as they became pregnant. Not all combat jobs are strenuous nor exposed to danger. The average American woman is pregnant for a very small portion of her life, and some never become pregnant. The services must guard against establishing policies that limit women to noncombat specialties merely because women may become pregnant.

Similarly, there is still ignorance about women's menstruation and a certain amount of mystery attached to this normal biological process. For most women, menstruation does not change their behavior very much. Also, studies have shown that women under stress or involved in strenuous physical activity may not menstruate at all. Nurses have served in wartime under primitive and unsanitary conditions, and questions have never been raised about negative effects of menstruating on their performance. This is a classic sexist myth that men have used for years to keep women locked out of combat roles.

Despite evidence to the contrary, some people hold the view that women do not handle stress as well as men.

Air Force Titan missile combat crew commander performs her duties at McConnell Air Force Base, Kansas (Official US Air Force photo).

Again, the satisfactory accomplishments of American nurses in combat and as prisoners of war in World War II have been recorded in history.

Finally, the idea that the integration of females in our military forces will have devastating consequences on unit cohesion is untested. We simply do not know whether the likelihood of success in battle would be diminished by using female "buddies." Studies involving women in similar positions such as police work will be useful in this debate, which is likely to continue until women actually find themselves in combat situations.

But now, despite old-fashioned views, myths, and out-moded social customs, the floodgates have been opened and new opportunities for military women are reported almost weekly. The woman mentioned earlier who was selected to command a flight squadron said, "Two weeks before I entered flight training, people were saying women would never be Navy pilots . . . . I look at young women entering the program today who are being told that they can't do many things, as I was told I can't do many things, and there's no doubt in my mind that they may well find themselves leading men and women into combat."

Many people believe that technology has made the combat exclusion law an outdated relic. General Holm asks, "What really are the differences between the person who repairs a torpedo on a submarine support ship and the one who fires it? Between the person who fuels a bomber and the one who flies it? Between the air traffic controller on shore and the one on an aircraft carrier? . . . Between the person who launches a Titan ICBM from a silo and the one who launches a Cruise missile from an airplane? Between the skipper of a Coast Guard cutter and the skipper of a Navy destroyer?"

Charles Moskos, a noted sociologist, states that the role of women in combat is unclear. "They do not serve in direct combat units, but some hold assignments almost sure to get them killed in wartime." The real issue, he says, is the prospect of compelling women to fight, to join combat units if the need arises, just as male recruits may be assigned to the infantry against their will.

Modern warriors find it increasingly difficult to enforce the many restrictions on women in combat. Army and Air Force women participated in the Grenada invasion in October 1983, and women Air Force and Navy pilots flew military planes in connection with the American action against Libya in 1986.

It has been reported that female air crews flew into Grenada during the first few hours of the invasion. Major General William J. Mall, who commanded the first wave of aircraft to attack, told the Los Angeles *Times*, "There was some hostile fire, but not in the immediate vicinity. There was some [fire] at one end of the airport." He explained that the staff knew that women would be arriving on the transports, but no one brought up US Code 10 Section 8529, which prohibits female members of the Air Force from duty in aircraft engaged in combat missions.

One member of the planning staff said that there was no time to hand-pick the crews. Planes were diverted from wherever they happened to be in the world and sent to Grenada. He reasoned that, "To have excluded an aircraft because there was a woman on board would have lessened our response and reduced our effectiveness."

Approximately 170 women soldiers were a part of the United States Forces team in the Grenada operation. Military policewomen were guards at the "prisoner of war" camp, drove jeep patrols, and served on roadblock and checkpoint duty. Other women soldiers were intelligence specialists, prisoner of war interrogators, and stevedores who documented and loaded into aircraft and a ship the Cuban/Soviet weapons and ammunition. The woman lieutenant whose platoon was in charge of telephone communications was called "Ma Bell." Truck drivers, helicopter crews, and one helicopter pilot were there along with administrative and postal clerks, laundry and bath specialists, and medical personnel. A woman ordnance captain was in charge of detonating unexploded bombs, grenades, and unserviceable US and Cuban ammunition left in Grenada.

In November 1983 the Coast Guard sent three ships to patrol Grenada waters after the departure of the Navy. The Executive Officer of the largest of the ships, a buoy tender, was a woman ensign.

However, this operation had its problems. Four of the military policewomen were sent to Grenada from Ft. Bragg on October 29, only to be returned when the field commander decided it was too dangerous. This was in violation of the Army's Direct Combat Probability Code. The women were subsequently returned, and in a letter to Representative Patricia Schroeder of Colorado, Secretary of Defense Weinberger explained:

"The problem illustrated in Grenada was that at a time of unit deployment into a crisis area—a critical time for unit integrity, cohesiveness, esprit, and teamwork—competent, valuable, experienced soldiers, many times key members of the team, are not deployed because they happen to be women. The unit suffers because it is going into an operation without all the members of the team; the individual soldier suffers a direct morale problem because it is a slap at her professionalism or her dedication to a military enlistment or career. One does not join the military to lead a sheltered life and be protected. A soldier knows the potential dangers of a military life. To be told, essentially: 'It doesn't matter how much you train, it doesn't matter how well you know your job, it doesn't matter how many months you have spent in the field surviving, it doesn't matter how important you are to your unit—you will not be deployed with your unit because you are a woman' is an affront. It is an affront not just to women but to all professional soldiers, men and women, who train together in peacetime to build a strong team. Furthermore, it is dangerous to the mission of the United States Army; and, finally, it is a waste of tax money—yours and mine—to train someone at great expense to our government and not to utilize her."

At the beginning of this chapter we called the combat issue a dilemma. The picture has improved remarkably, but women will not be true and equal partners in defense until society, Congress, and the services resolve the problem in a climate free of bias and myths.

CHAPTER 6

DACOWITS (Defense Advisory Committee on Women in the Services)

The DACOWITS (Defense Advisory Committee on Women in the Services) has been coping with sexism in the armed forces since 1951. It would be impossible to trace the progress of women in the military without examining the leadership of these civilian men and women.

The DACOWITS was established in 1951 by Secretary of Defense George C. Marshall to give assistance and advice on policies and matters relating to women in the military service. Mrs. Anna Rosenberg, Assistant Secretary of Defense for Manpower, Personnel, and Reserve, served as chairman-hostess for the original group of women who were from business, education, and civic enterprise and represented all parts of the country.

Reflecting the problems and culture of the era, the Committee's objectives were:

- to inform the public of the need for women in the services;
- to emphasize to parents the responsibilities assumed by the military departments to provide for the welfare of women in the services; and
- to accelerate the recruitment of women, stressing both quality and quantity.

Contrast those objectives with the ones established by the Secretary of Defense in April 1987:

- to provide the Secretary of Defense assistance and advice on matters relating to women in the services and recommend measures to ensure effective utilization of the capabilities of these women. Members may serve on task forces covering issues such as recruitment, training, utilization, housing, health, and welfare;
- to interpret to the public the need for the role of women as an integral part of the armed forces, encourage public acceptance of military service as a citizenship responsibility and as a career field for qualified women, and provide a link between the armed forces and the civilian community.

Its first major undertaking, and in fact its reason for being, was the ambitious recruiting campaign during the Korean War years. During the Committee's formative year, it made fifteen recommendations, of which ten were carried out by the Defense Department.

The earliest focus was on recruiting more women and educating the public. Films and brochures were produced, often aimed at the parents of young women. The members visited military bases and were popular speakers at men's and women's civic clubs.

In 1955 they expanded their horizon and conducted a study of housing conditions for women officers. This was the beginning of a long-term commitment to improved living quarters for single women that eventually led to better quarters for single men as well. That year also saw the Committee recommend opening the Reserve Officers' Training Program (ROTC) to women.

In the next few years the DACOWITS took an active interest in legislation that affected women, their careers, and their veterans benefits. They backed the Career Incentive Bill for Nurses and Medical Specialists and the Women's Army Auxiliary Corps bill, which gave veteran status to early members of the WAAC who had been shamefully left out. The DACOWITS recommended legislation and policy changes to remove the grade restrictions on women officers, thereby allowing women to rise to the ranks of general and admiral; to discontinue the unfair system of computing housing allowances when both husband and wife are military; to allow civilian husbands of service women to use the Post Exchange, a privilege always extended to civilian wives of male servicemen; and to permit women with dependents under the age of eighteen to remain in the service.

On the recommendation of the Committee a working

Captain Susan D. Rogers, one of the first group of women to enter undergraduate pilot training in 1976, makes a pre-flight check on the wheel of her T-38 trainer (US Air Force photo).

group of senior officers was formed in 1966 to study the policies and objectives regarding the use of women in a military capacity. Colonel Jeanne Holm, who later became the Air Force's first woman general and the universally acknowledged authority on military women, was the chair-

man. In later years, General Holm was appointed to the
DACOWITS.

The DACOWITS continued its interest in housing,
especially for young and single members, because of the
large number of women in those categories. Living conditions were considered of prime importance if the services
were to attract and retain high-quality men and women.
They advocated the elimination of open bay barracks,
minimum standards of occupancy (no more than two to a
room), and the construction of multipurpose housing units
where bachelors and families could live. This a wonderful
example of how everyone's status is improved when one
group speaks up for its rights. Today young and single
servicemen enjoy far better living standards because of the
interest the DACOWITS took in the environment of
servicewomen.

In 1974, for the first time, the DACOWITS directly
addressed the issue of women in the service academies and
made a strong recommendation favoring their admission.
The resulting law was signed by President Gerald R. Ford
in 1975 and was effective the following July.

With the end of the draft, use of women moved to the
top of the agenda, and in 1974 the Committee recommended legislation to remove restrictions on training of
service women for combat-related jobs. They also recommended that the Air Force revise its rules to permit
women to become pilots. In August 1976 the first Air Force
women were admitted to flight training at Lackland Air
Force Base.

The DACOWITS supported an amendment to US Code
10 to allow women to serve in a temporary duty capacity on
ships not expected to be engaged in combat missions and
allow them in a permanent status on hospital, transport,
and other vessels not involved in combat duty. As a result,

the first Navy women reported for sea duty on the repair ship USS *Vulcan* on November 1, 1978.

In 1986, as the DACOWITS celebrated its 35th anniversary, they took justifiable pride in the progress made by military women, as evidenced by events such as these:

- Army female helicopter pilots and military police participated in the airlift of Honduran troops during the Nicaraguan incursion.
- Air Force women served in flight crews during aerial refueling operations during the US action over Libya.
- 320 billets aboard Navy Military Sealift Command ships were opened to women.
- Cadet Terrie Ann McLaughlin graduated first in her class from the United States Air Force Academy.

Current issues of concern to the DACOWITS are:

- utilization and deployment of military women in light of existing legally imposed combat restrictions;
- impact of the declining military eligible male population on the role of women;
- assurance of adequate housing, health care, and child care for all servicepeople;
- elimination of all discrimination and sexual harassment;
- establishment of equal minimum admission standards to all services;
- enhancement of the role of women in the Reserve Forces and the National Guard.

Because of talent, experience, influence, and prestige coupled with great insight and tenacity, the DACOWITS has been an extraordinarily effective advocate for military women. There are few of us who would not recognize many of the names found on the roster of present-day and former members:

Marcia M. Carlucci, wife of the Secretary of Defense; Mrs. Milton Berle; Mrs. Anna Chennault; Mrs. Elinor Guggenheimer of New York; Paula F. Hawkins of Florida; Mrs. Oveta Culp Hobby, first Director of the WAC; Major General Jeanne M. Holm, first woman to be promoted to that two-star rank; Representative Barbara Marumoto of Hawaii; Mrs. Sarah McClendon, journalist and World War II WAC; Justice Sandra Day O'Connor, first woman appointed to the Supreme Court; Martha Raye, movie star famous for her work entertaining American troops in the war zone; Mrs. Mary C. Rockefeller; Mrs. Thomas W. Streeter, first Director of the Women Marine Reserve in World War II; and Mrs. Arthur Hays Sulzberger of the New York *Times* family.

CHAPTER 7

Vive la différence

A popular expression not long ago was, "Men are men. Women are women. *Vive la différence!*" Well, it's *la différence* that has caused so many crusty old soldiers to tremble at the sight of a woman in uniform. It's safe to say that some rather simple and ordinary issues have been the source of greater irritation, conflict, and misunderstanding over the years than whether or not women should serve in combat.

Women brought a new set of values and needs to the military. Some were as trivial as wanting to wear earrings in uniform; others were as important as the requirement for quality gynecological care. Soldiers, sailors, and marines who had proved fearless in battle retreated in haste from discussions of menstrual tampons and feminine underclothes. Military women sometimes wondered if the men they worked with had mothers, wives, and daughters.

When women first joined the services in large numbers, there was much haggling over whether they should be issued bras, girdles, pajamas, and bathrobes. Some argued that since the men did not receive these items, it would be wrong to give them to women. Others reasoned that pajamas and robes were needed because women were more modest than men and not accustomed to walking around undressed in a group living situation. In the end, it was decided to give the women an allowance to buy their own underclothing.

Gender-related problems affect a person's everyday life in hundreds of ways that may seem small and insignificant but, in fact, are important to self-image, morale, and physical and emotional well-being. In this chapter we will discuss appearance, uniforms, health care, living quarters, fitness standards, discipline, fraternization, recreation, entertainment, and the Code of Conduct.

Military men and women do not find it easy to agree on these issues. In fact, there is controversy even among the women on what they themselves should expect. Some problems have been resolved over time, and today all of them are openly discussed and debated. The DACOWITS and even the Congress have prodded military leaders to reexamine policies and attitudes to weed out unfair bias and to make appropriate accommodations that recognize that women are a permanent part of the United States military forces. The quality of life enjoyed or endured by servicewomen is a topic of great interest and is currently receiving the serious attention it deserves.

APPEARANCE

For too many years military men had been overly concerned and thoroughly confused about how military women should look. Senior leaders often ranked attractiveness along-side competence as desirable attributes when recruiting

women. Any indication of masculinity was unacceptable;
athletic women were suspected of being lesbians. Yet,
decidedly feminine women were apt to be seen as weak
and unprofessional.

In 1964 a Marine Corps study group examining the role
of women said, "Women Marines must be the smallest
group of women in the military service. In accordance with
the Commandant's desire, they must also be the most
attractive and useful women in the four line services."
Using the same line of thinking, in 1966 the Air Force
Chief of Staff ordered the Recruiting Service to get "better-
looking WAF."

Senior military women were as guilty as men in this
regard. It was understood and the message was unmistak-
able from the top down that servicewomen were expected
to look feminine in uniform and civilian clothing. Well into
the 1960s regulations demanded strict adherence to an old-
fashioned model of the ideal woman. Slacks were seldom
considered appropriate attire. Women Marines could not
wear shorts unless participating in an active sport. Even
then, they were ordered to cover the shorts with a shirt en
route from the barracks to the playing field.

Hairstyles were strictly prescribed. Hair had to be neat
and feminine—never too short for fear of giving a masculine
appearance. Makeup, at least lipstick, was often required.
Military women were given instructions in the proper
application of cosmetics.

The most attractive women were routinely assigned to
the most visible and often most desirable jobs. To be pretty
and well groomed were the first requirements for positions
in the public eye or in the Pentagon.

In fairness, it must be said that this preoccupation with
attractiveness is not unique to the military. It often happens
that when women break new ground and enter formerly

male territory, some men make unkind and irrelevant remarks about physical appearance. Homely and even average-looking women are suspected of looking for a husband; athletic women are suspected of being gay; and pretty women are asked over and over why they decided to become an engineer, a doctor, a construction worker, a scientist, a firefighter, or a policewoman.

Today women are no longer such an oddity in the military that they must apologize for a less than perfect face or figure. The armed forces have matured, and women's role in defense has become too vital for such trivial and superficial concerns.

Many but not all women are assigned to nontraditional jobs that make it necessary to wear sturdy, practical clothing; hair gets wet in the rain; fingernails get dirty; manicures are ruined; military women perspire. The expectations are much more realistic. Being pretty is not an acceptable substitute for competence.

Today the military is probably no better or worse than other predominantly male organizations in judging women by standards based upon appearance and notions of femininity. This issue has not disappeared entirely, but it is no longer of great importance and should not concern the modern young woman considering a career in the armed forces.

A related point is the question of size. Generally, women have been held to stricter height and weight standards than men. Whereas some argue that women are not big enough or strong enough for military duty, others write policies that screen out tall, heavy female applicants. The armed forces have been accused of not accepting women who are too large to look well in uniform.

Women who meet minimum standards for firefighters are considered too tall and heavy for the services. In the past it was argued that it was too difficult to provide suitable uniforms for women who were larger than average; but it is more likely that heavy women did not fit someone's idea of womanhood.

This is not a question of fatness, which we know is unhealthy. This is a matter of large-boned, heavily muscled women who are fit and strong. In most of the services men are allowed to be taller and to have a larger proportion of body fat than women. The Air Force is moving the fastest in narrowing the gap in male/female physical standards for enlistment. All the armed services have been forced to confront this issue. The DACOWITS and some members of Congress have taken an interest in it.

UNIFORMS

Dressing military women has been a problem from the beginning and remains so to this day. Should women's uniforms mimic the men's with a decidedly masculine look? Can they be practical, yet feminine? How closely should they fit the body? Will someone please design a hat that does not destroy a woman's hairstyle? Can stockings have seams? Should waistlines be pinched or should jackets be boxy? Blouses tucked in or worn over the skirtband? Earrings? Umbrellas? High heels? Flat heels? Pointed-toe shoes? Long skirts? Mini skirts? Skirts just below the knee? Full slips? Half slips? No slips? Bras? Wear a sword? What about maternity uniforms for the office? for formal dress occasions? for field exercises?

The world of fashion dictates more changes for women than for men. That causes a certain amount of havoc in women's uniform regulations and styles. It is irritating and expensive, especially if the uniforms are tailor-made.

Besides the matter of style, there has always been a problem with adequate supplies of required items. Because there are so few women at most military posts, the store-keepers too often do not stock a range of styles and sizes.

In recent years uniform problems have taken a decidedly different turn. The questions now are more often ones of function and practicality than of fashion. When women were first integrated into male training units they needed field boots. None were designed specifically for women, so they were issued men's boots, which were too large and did not adequately support women's smaller ankles and legs. The result was predictable—too many women on crutches.

Women did not have sturdy field clothing and were issued men's jungle fatigues. Since women's bodies are proportioned differently, fitting was a problem.

A more serious uniform situation arose during the Vietnam War when it became obvious that military women assigned to Southeast Asia did not have adequate clothing. After the Tet Offensive some commanders ordered all personnel to wear field uniforms. Army women changed from the green cord two-piece skirted uniform and high heels to the olive green trousers and men's shirt worn with high-top laced boots and a baseball style cap.

The WAC Director in Washington was adamantly opposed to the outfit because photos in the newspapers showed the young women looking like combat troops. She feared that it lowered the prestige of the military women and cause their parents to worry. In her view it did not present a neat and feminine appearance. When she insisted that the women get back into skirts, one WAC major wrote from Vietnam that the women resisted because they felt " . . . that the WACs are in Vietnam to do a job and not to improve the morale of the male troops."

The Air Force women serving in Vietnam were issued a more ladylike outfit that consisted of a light blue overblouse, navy wrap skirt or slacks worn with sneakers, and no hat. General Jeanne Holm, onetime Director of the WAF, says, " . . . it was attractive yet functional—except when diving for cover in a bunker, and then it was a mess."

Women marines heading for Vietnam were advised by the Corps to take four to six pairs of high heels to wear with the uniform because the streets were hard on shoes and repair service was unsatisfactory. The "Information on Saigon" booklet provided each woman before leaving the United States advised, " . . . bring a dozen sets of heel lifts . . . Heels can easily be extracted with a pair of pliers and new ones inserted with little difficulty."

In war-torn Vietnam nylon stockings were a luxury. They were difficult to find and hard to maintain. Women of some services were excused from wearing them when in uniform. This privilege was not extended to the women marines, who were expected to look neat and feminine under all conditions.

Twenty years later military women still have similar complaints. In 1987 Navy women were asked to rate their uniforms on these factors: comfort, safety, appearance, quality of material, availability in the local Navy Exchange, sizing standards, quality and fit of the safety shoe. Eighty-five percent of those surveyed said they were dissatisfied in at least one of the above categories. The main complaints were aimed at availability, sizing, and quality of material.

To make matters worse, aboard ship the female senior officer's hat could easily be lost overboard because the chin strap is not functional.

Just as in the 1960s there were no field boots for the women in Vietnam, in the 1980s there are no safety shoes made specifically for women. The men's shoes still do not fit, and it is common for women to wear two and even three pairs of socks with them.

It was learned that women's uniforms were often more expensive than men's. For example, in the 1987 Navy study the following cost comparisons were reported:

	Men	Women
Enlisted cap	$ 2.65	$ 21.70
Junior officer hat	$25.70	$ 59.00
Senior officer hat	$44.20	$107.50

Perhaps no other piece of military clothing tells how far women have come than the camouflage maternity uniform. In 1989 the Marines joined the Army and Air Force and became the last of the services to make maternity "cammies" available. These uniforms have become a necessity in today's military where many women, even in advanced pregnancy, perform duties that require uniforms that are comfortable and easy to maintain. Pregnant women receive a supplemental maternity allowance to purchase maternity uniforms.

Wearing a smart, well-fitted uniform is a source of great pride for members of the armed forces. Women certainly want to look as well as the men, but very often they are frustrated by the fit, design, quality, appropriateness, and availability of uniforms. It seems that less progress has been made in this seemingly simple area than in the more complicated realm of job assignments.

MEDICAL CARE

Free medical care is one of the most important and attractive benefits offered to new enlistees. It has the dual purpose of ensuring a healthy and fit army and at the same time guaranteeing a trained and ready medical corps. Until recently, military health care focused on active duty men and their families.

For a variety of reasons little thought was given to the needs of female military members. Because until the 1970s women who were pregnant had to leave the service— whether married or single—there was no requirement for obstetrical care. Many young women left military service soon after marriage because the policies of the time made family life nearly impossible. The result was fewer middle-aged women, thus reducing the need for gynecological care. Until recently, because women did not serve on board ships or in remote locations, female care was not provided.

Abortion and birth control were seldom discussed in public. A married servicewoman would have asked for birth control pills in her role as someone's wife and not in her own right. A single woman would not have asked at all; she would have been branded a "loose woman." Today, birth control pills and devices are available to married and single servicewomen, but they can obtain only a thirty-day supply each time whereas wives of servicemen are given prescriptions for a six-month supply.

Most military hospitals and clinics have a policy of treating active duty members before wives and children of servicemen so that uniformed persons lose as little work time as possible. Because no one foresaw that female soldiers would be seeing obstetricians, gynecologists, and pediatricians, the policy of priority for servicewomen was nonexistent. As recently as 1987 women in uniform complained that they were often treated after dependents, thereby losing valuable time on the job.

As a result of the dissatisfaction expressed by many women, the Defense Department conducted a survey in 1987 to find out whether military hospitals, clinics, and medical personnel were meeting the needs of women in the service.

Fifteen thousand women in the Army, Navy, Air Force, and Marine Corps received an 86-item questionnaire covering such topics as access to health care, pregnancy, contraception, and awareness of disease-prevention techniques.

They were asked questions such as, "How long did you have to wait for your last appointment with a gynecologist?" and "Have you ever experienced harassment when seeking a contraceptive or birth control device from a military medical treatment facility?" Another question was whether they had been taught to do their own breast examination since coming on active duty.

The results showed that most active duty women were satisfied with the treatment they received at military hospitals and clinics, but many believed that service wives had higher priority for appointments. Since about 64 percent responded that they were generally satisfied, one official said the survey demonstrated that " . . . military obstetrics and gynecology care is not the 'disaster' that some critics say it is."

Air Force women were happier about many aspects of their care than were women in the other services. Coast Guard women consistently showed greater dissatisfaction.

Defense officials said that the issue of servicewomen having to wait until after dependents were served may have resulted from poor communication between hospital staff and active duty women who failed to identify themselves as such when telephoning for appointments.

Everyone was pleased that 82 percent of pregnant active duty women surveyed reported regular prenatal OB/GYN appointments with military doctors. Considering the country's crisis in prenatal care, this was good news and bodes well for military children.

Women felt that one positive result of the survey was the heightened awareness of the problems. Captain Shelley Mitchell, a woman Marine company commander, said that the questionnaire made her realize that she was not alone. In her words, "You know if there's a question about something on the questionnaire that means someone else has reported something about it." She added her own view that ". . . women suffer most from the prejudiced attitudes of some medical personnel. If a woman complains of pain or illness, sometimes they don't take the woman seriously . . . It may be ignored because some feel women are 'wimpy' and it's not really a bad problem." In fact, some women do have a low tolerance for pain, Mitchell said. But other women are so tough that ". . . they have to be almost dying before they would complain."

The following table taken from the 1989 DoD Women's Health Care Survey preliminary report illustrates the servicewomen's view of care.

	Army	Navy	Marine Corps	Air Force	Coast Guard
Very Satisfied	13%	14%	13%	21%	21%
Satisfied	46%	50%	48%	51%	44%
Neutral	23%	17%	18%	16%	23%
Dissatisfied	15%	15%	16%	10%	14%
Very Dissatisfied	3%	4%	5%	2%	6%

LIVING QUARTERS

Contemporary Bachelor Enlisted Quarters (BEQs) and Bachelor Officers' Quarters (BOQs) bear no resemblance to the barracks that were standard until the 1970s. On many military bases, but not all, servicemen and women are housed in motel-style buildings with two or three persons to a room. There are lounges, laundries, and dining halls. These comfortable surroundings cannot be truly appreciated without some understanding of the problems encountered by the first military women.

Early in World War II the senior women officers recognized that the physical layout of barracks did not suit women. Psychologically, emotionally, and culturally, women could not adapt to the open squad rooms where bunks were lined up one after another with no partitions for privacy. The atmosphere was stark and not at all homelike. There was no place to receive guests, make snacks, hang wet laundry, or spend quiet time alone. The walls were most likely painted gray or green and were bare except for official items such as fire regulations, security instructions, and exit signs.

Curtains, bedspreads, colorful towels, small rugs, dressers with drawers, photos, and stuffed animals are a few examples of the changes women eventually brought to barracks living. Females have traditionally been vitally concerned with their home. They generally spend more time in their quarters and have some needs that differ from the average male's.

Guest lounges were unique to women's quarters. One room, usually furnished with comfortable furniture, a record player and later a TV, was set aside for greeting and entertaining dates. The dress code in the guest lounge was very strict: Men and women had to wear full uniform or comparable civilian clothing. For the women, sportswear, shorts, or slacks were definitely not permitted.

Eventually, but not without a fight, sewing rooms, hair dryers, and small kitchens appeared. Shower curtains were put up, and doors were installed on toilet stalls. It was left to women commanders to explain that women did not send personal clothing to commercial laundries and therefore needed more washing machines, dryers, and ironing boards than government specifications allowed. These were the issues of the 1940s and '50s.

Yet in 1987 the DACOWITS expressed concern that minimum housing standards and the unique needs for privacy and security for women were still not being met. At their semiannual meeting, they asked the Defense Department to explain standards for bachelor enlisted housing by rank and gender. Specifically, they requested information about room and building security, lounges, cooking facilities, room sizes, privacy, fans/air conditioning, curtains/blinds, mirrors, and other furnishings.

As new living quarters are built around the world for our military men and women, many of the problems will fade, but there will probably never be complete agreement on what constitutes adequate housing for women. One thing is pretty sure: Men's barracks have been improved and better equipped as a result of women's insistence on nicer living conditions.

PHYSICAL FITNESS STANDARDS

The purpose of strict fitness standards is simple: The armed forces must be ready and strong. The only question, then, is whether men and women should be measured by identical criteria. Until the 1960s not much thought was given to the strength of females. Looking back, that is surprising, because during the era of World War II and Korea women filled many strenuous assignments—far more than in the 1960s and '70s.

Physical training was designed to keep the women trim. Most of the time the workouts mimicked high school gym classes, which were usually a series of stretching exercises and a medium-paced game such as volleyball or softball. Once again, the military merely reflected the current cultural beliefs of the country. In *Women in the Military* General Holm writes, "Ruggedness was certainly not the name of the game—good health and appearance were."

Each service has a different philosophy of physical conditioning, based on its role. The Army and Marine Corps put the greatest emphasis on strength and endurance because of their infantry combat mission. The Navy, after basic training, tends to concentrate on "fitness for life," with education about smoking, drug and alcohol abuse, stress management, exercise, nutrition, and accident prevention.

In the end, what each service wants is people who can do their job. There is a trend to use physical-strength standards based on the requirements of assigned tasks, but the idea has not yet been successfully implemented. At one time the Army planned to adopt the Department of Labor job categories: sedentary, light, medium, heavy, and very heavy. For example: "medium" work consisted of lifting weights up to 50 pounds and frequently lifting 25 pounds.

After the criteria were developed and the Army jobs were classified, it was found that 92 percent of the women were ineligible for 76 percent of all jobs. But in 1982 more than half of the Army's women were doing those same jobs.

Marines have always maintained separate physical fitness standards and testing for men and women. In their own words, "Presumably, this can be attributed to traditional perceptions of the proper employment and capabilities of women, within or without the Marine Corps."

We will probably hear more debate on this issue. No one seems to disagree with the premise that men and women differ physiologically. Generally, women have less upper-body strength. But should standards be based on gender, or should one set of standards be used for everyone, thereby allowing stronger women to be eligible for "heavy" jobs? Should jobs be open to all based on physical strength requirements rather than gender? Should men and women train together? There is evidence that physical training sessions not only build strength but also team spirit and comradeship, two highly valued qualities in the armed forces.

DISCIPLINE

Statistics show that the disciplinary offense rate for military women is much less than for men. No one argues that fact, but they will dispute the reason why. Is it because women are better behaved? Are women more carefully screened for enlistment? Are the disciplinary standards different for men and women? Or are women treated more leniently?

The answer is not easy to find. Enlistment criteria were traditionally more stringent for women, and to some extent that is still a factor. Military women were expected to follow all the rules—military, social, and cultural. Single women who became pregnant were automatically discharged even if they miscarried, because the military could not "condone" an unwed pregnancy or a "dilution of standards set for women . . . " In 1972, when a woman charged that the Navy maintained one standard of behavior for women and another for men, the Deputy Chief of Naval Personnel stated that he "did not accept the rationale that men and women should be held to a single standard of morality." So it can be argued that women were treated more harshly than men.

On the other hand, for many years the armed forces were apt to discharge women who were disciplinary problems, whereas men were jailed for similar offenses. Presumably this was because men had a legal obligation to serve whereas women were volunteers. It was generally accepted that if a woman was a problem it was better to get rid of her. Men, however, could not be allowed to escape military service by committing minor infractions.

Military men have accused women of taking advantage of their sex to gain preferential treatment. In 1988 the Marine Corps specifically studied this allegation and concluded that although the punishment rates were lower for women, the statistics did not "support the perception of leniency toward women in disciplinary actions."

The Navy, at the same time, found that the court-martial rate for women was one-fifth that of men. Women were first confined in Navy jails, called brigs, in 1983. Previously, they were discharged for minor offenses or sent to federal women's prisons if they had committed serious civil crimes. Since 1983 the brig population of women has remained steady at about 1 percent of the total prisoners per year.

When commanding officers of ships with mixed crews were surveyed they said, "... as women accept roles traditionally filled by men, it should be neither surprising nor alarming that women's behavior—even when inappropriate—is at times similar to that of their male shipmates." This acknowledgment signals a new spirit of equality between the sexes. It seems that sailors are on the way to being judged by their behavior regardless of their gender. Women have even earned the right to go to the brig!

In truth, women in civilian occupations are faced with the same issues. They feel they are expected to uphold a different set of standards and are more apt to be fired than a man in the same circumstances. Civilian men believe that women have an advantage because of sex and are treated with leniency. Probably there is some basis for all these fears. Professional women, military and civilian, cannot succeed unless they are aware of these perceptions and learn to deal with them. Equality means the same pay, the same opportunities, and the same punishment.

FRATERNIZATION

Historically, the military has espoused a policy of non-fraternization. This concept goes back to our European roots, but in all that time few have been able to satisfactorily explain or enforce it.

The *American Heritage Dictionary* defines the word "fraternize" as "To associate with others in a brotherly or congenial way." When there were no women in the armed forces it was a simple idea: Male officers could not socialize with enlisted men. Some argue that the custom evolved from a notion of a class system because officers were from the gentry and enlisted men were from the lower classes. Others say that it is based on the idea of fairness because officers have control over enlisted promotions and assignments.

The Officer's Guide in the 1940s stated, ". . . officers and men have not generally associated together in mutual social activities. No officer could violate this ancient custom with one or two men of his command and convince the others of his unswerving impartiality. The soldier does not need or desire the social companionship of officers." While it was always an irksome rule, it did not cause much real controversy until World War II, when large numbers of enlisted men joined the officer ranks and even worse, women arrived both as officers and enlisted.

We shall focus on the male-female aspect here. American men and women are accustomed to dating outside their own status. Especially, it has been a fact of life that women date older men with more education, more money, and better jobs. This social custom, so long accepted in civilian life, was challenged by the equally entrenched military custom of nonfraternization.

Since World War II most services refused to publish the policy with all its specifics because it was such a nightmare to explain. It was difficult for civilians to understand anything so "snobbish." It was impossible for commanders to enforce with any semblance of impartiality.

For many years a large majority of men turned their heads when they knew of a male officer dating an enlisted woman. Just the opposite happened when the situation was reversed: A female officer dating an enlisted was regarded with some disdain. If a mixed relationship became too obvious, it usually led to one of the parties leaving the service—in almost all cases the woman left. Enlisted men resented enlisted women dating officers and equally resented being told that they were not good enough to date female officers. Much of the same attitude lingers today.

It is even more perplexing in an enlisted military marriage when one partner is promoted to the officer ranks. This situation calls for great patience, common sense, and discretion because each partner presumably has a different set of friends. Military housing and social functions are generally clearly segregated by rank. It affects not only the couple but all their friends, coworkers, children and the children's friends.

In 1942, to no avail, WAC Director Colonel Hobby urged the Army to adopt a more relaxed rule and to spell out the details. She was consistently overruled. Perhaps it was thought that the issue would eventually resolve itself, but it has not done so.

To magnify the problem, some services define fraternization as social interaction between *all* seniors and subordinates. Therefore, privates cannot associate with non-commissioned officers; non-commissioned officers cannot associate with staff non-commissioned officers; junior officers cannot associate with field grade officers and so it goes.

In 1978 a West Point graduate resigned because he had married an enlisted woman and the Army would not accept it. In 1987 the Marine Corps brought court-martial charges against a woman Navy dentist who married a Marine corporal. If she had been convicted, she would have faced a dishonorable discharge, two years in prison, and loss of her license to practice dentistry. Her husband left the service, and shortly thereafter the charges were dropped. In 1988 a Marine lieutenant was sentenced to dismissal from the service after being found guilty of fraternization and adultery with an enlisted woman.

Before the 100th Congress adjourned, Representatives Beverly Byron and Rod Chandler sought a review of the rules on military fraternization. They hoped to prompt Pentagon officials to evaluate policies and make needed changes. Congresswoman Byron said, "I am convinced that people in the military don't know what the standard is. Either we need to say we do not tolerate fraternization under any circumstances, or we need to find a way to accommodate legitimate dating where there is no chain-of-command problem."

The services began to work on defining their policies. The Navy, for example, wrote instructions that said in part that fraternization is " . . . any personal relationship between and officer and enlisted member that does not respect differences in rank where a senior-subordinate relationship exists." The Navy personnel chief said he expected people to exercise good judgment and recognize that there are human beings on the other end of the policy.

The Coast Guard published its first-ever fraternization policy and went to great lengths to give examples of proper and improper relationships. It was applicable to all, regardless of gender, and leaned heavily on the appearance of impartiality. For example, it stated that a commanding officer and a chief may go fishing together, but it would be wrong to do so while everyone else was at work. It banned social relationships between instructors and students. A key point in judging the propriety of any relationship between members of different rank according to the Coast Guard is the authority or influence one member exercises over the other.

As you can see, fraternization is still a fuzzy issue. Most people agree, however, that there is need for a nonfraternization policy of some sort because without it discipline, morale, and management would be in chaos. Civilian organizations face the same problem, although they are less candid about it. There are cases of women leaving jobs because they date the boss, and even cases of employees being forced out of a company because they flaunted a personal relationship that caused hard feelings among coworkers. The military is more open about it, but all working women must be aware of the risks of dating or marrying superiors, supervisors, or subordinates.

RECREATION AND ENTERTAINMENT

Recreational facilities and programs on military posts and stations were planned by men for men. There has always been heavy emphasis on sports, gyms, and hobby shops. The increase of military women has had an impact and caused the services to modify their thinking and their spending.

Fitness equipment more suitable for women is being

purchased, and locker rooms are being refurnished. Aerobics classes attract both sexes. In fact, the improved atmosphere of some gyms has resulted in greater use by both sexes.

While current budgets routinely consider the recreational needs of women, both military and family members, off-duty activities for females will probably never match those provided for men. The difference lies in the numbers. Overseas the problem is usually greater, because there are fewer women. Interestingly, in 1987 women overseas said they would be satisfied to have a kitchen-like facility where they could get together in a relaxed homelike atmosphere.

The on-post clubs where service members congregate to eat, drink, and socialize have been a source of dissatisfaction for women. Again, it's the ratio. The women often feel uncomfortable when they are so greatly outnumbered. It's hard to find a quiet place just to visit and talk. There is a perception that women go to the clubs for the company of men, and that can lead to hard feelings when men are rejected and discomfort to women who are repeatedly interrupted by unwelcome males.

A more serious situation came to light when members of the DACOWITS toured bases in the Pacific area in 1987 and found that the entertainment in military clubs was sexually oriented, offensive, and degrading. The situation was intolerable for servicewomen, who had no other place to go for relaxation.

This is a clear example of lack of sensitivity. Regulations had been on the books for years barring such an unwholesome environment, but few paid attention or gave any thought to the women who would be using the clubs. Congress and the Defense Department brought immediate pressure on the services to improve the conditions for women, both females in the military and spouses of members. Commanders were reminded of their responsibility to ensure that events on their posts met acceptable standards of discretion, modesty, and good taste.

Since that incident, studies have been made, women have been surveyed, and regulations have been revised to provide clearer standards.

On the plus side, women can expect to find a great variety of recreational and entertainment facilities on the average post—perhaps more than at home. And they will be far less expensive. But there can be no doubt that the balance is in favor of the men.

Women who serve in foreign countries, especially in the Third World, should be prepared to face some culturally imposed restrictions. In these areas, where American women cannot move about with the same freedom as at home, it is essential that the command be even more sensitive to their after-hours needs. It is still a man's world in many countries.

CODE OF CONDUCT

The Defense Department has made a concerted and successful effort to rid its directives and correspondence of gender-specific terminology. But sometimes the most obvious escapes even the best intentions.

Since 1955 every American servicewoman signed a Code of Conduct beginning with the words, "I am an American fighting man." The code was developed after the Korean War to govern the conduct of military personnel taken captive by the enemy. It went on to state, "I will never surrender my men while they still have the means to resist . . ." and it ended with the vow, "I will never forget that I am an American fighting man."

In 1985 a woman sailor objected and said she did not consider herself to be any kind of a man. Her letter took two years to reach the desk of the Secretary of Defense, who agreed with her. Accordingly, President Ronald Reagan signed an executive order on March 28, 1988, changing the wording to, "I am an American . . . I will never surrender the members of my command . . . I will never forget that I am an American."

SUMMARY

These are not all the gender issues that arise in the armed forces, or in the civilian work world, for that matter. In fact I have left the most important one—marriage and children —for a separate chapter. But I hoped to show how so many things we take for granted are based on our perceptions of what is masculine and what is feminine; on our culture; and on our real differences. As in so many situations, lack of sensitivity causes more problems than outright prejudice. Today women comprise 10 percent of the force. They have a visible presence and a louder voice. Because of the heightened expectations of the generation, women are less likely to silently bear the slights of the past. And the men who have working wives and daughters with meaningful careers are listening. It's an exciting time for women.

CHAPTER ◇ 8

A Woman's Place: Assignment Policies

T he opening of 24,000 new jobs to servicewomen in late 1988 marked perhaps the greatest break-through in gender-based assignment policies in the history of women in the armed forces. News headlines proclaimed that the Pentagon had redefined a woman's place. The shift was a reaction to a Defense Department report critical of the way the services decided where women could serve.

The move was dramatic, but another government study made to the Congress at about the same time argued that the number of new positions was far below the 880,000 that could have been opened up if the Department of Defense were more willing to make sweeping policy changes. "Unless the services have compelling reasons that they have not yet identified, all unrestricted jobs should be available to men and women on an equitable basis," said the report by the Government Accounting Office.

The disagreement between the two exemplifies the

stumbling blocks that have hampered women's progress in the past and persist today. The armed forces traditionally used the combat exclusion policy as a convenient rationale for limiting the role of women. But in practice they often barred women from thousands of opportunities for other than "combat risk" reasons.

The Navy policy for assigning berthing (sleeping) spaces aboard ships restricts women from serving aboard some ships. For example, if a berthing compartment accommodates forty people, the Navy assigns only forty women to that ship even if there are actually seventy jobs suitable for females.

Rather than open all noncombat-related pilot and navigator jobs to women, the Air Force limited the available positions to an artificial percentage of women it thought would be interested or could qualify.

The Army recruiters were given separate male and female job listings for noncombat-related jobs, and recruiters were not allowed to make gender substitutions when a job was filled in a particular gender category. The number of women allowed to join the Army is based on the Army's recruiting goals rather than on the number of interested and qualified women.

The Marine Corps limited the number of men and women in some noncombat-related job fields to a 50–50 balance to reflect the general makeup of the population. When there were not enough women to meet the quota, more men could be assigned to the field; but in the reverse situation the Marines did not allow women to exceed the 50 percent cutoff.

Although some did not feel that the services went far enough, most supporters of stronger roles for military women found the actions taken in 1988 heartening.

Under the new guideline, called the "risk rule," non-

Commander Deborah S. Gernes, Executive Officer of the destroyer tender USS *Cape Cod*, and a petty officer observe activity from the ship's bridge in Subic Bay, Philippines. Cmdr. Gernes was the first woman declared eligible to command a Navy ship (Official US Navy photo).

combat positions can be closed to women only if units with those jobs face a risk of getting their people involved in combat or captured that is equal to or greater than combat units in the same geographic area. The risk rule was a major effort to get the services to agree on which noncombat positions should be opened. It was also an attempt to systematize the standards used by the several branches of service. The study had found, for example, that reconnaissance aircraft were closed to women in the Navy but open to women in the Air Force because each applied a different rule of risk.

Major openings approved in the 1988 action included:

- Army—engineer combat support equipment companies; assignments in infantry, tank, air defense, and communications unit headquarters.
- Navy—combat logistics force ships (ammunition ships, oilers, and other support vessels), EP-3 aircraft, and underwater construction teams (frogmen).
- Air Force—high-altitude reconnaissance plane crews such as the SR-71 Blackbird, the TR-1, and the U-2 (spy planes).
- Marine Corps—security guard program (embassy guards) and Marine Corps security forces at designated locations.

The Coast Guard was not affected because it has been sending women to sea for more than a decade, imposing no restrictions whatever. Women in the Coast Guard serve in law enforcement and drug interdiction roles—the closest types of operations to combat that any seagoing forces are likely to experience in peacetime.

The new opportunities could correct another inequity, the lag in women's promotions. Women have achieved success in numerous career fields, but the highest levels of leadership have been out of reach for many. In all the services, the way to the top is through command: of a battalion, a ship, an air wing. Excluded from so many combat posts, women's military careers tend to peak at the middle ranks. About 18 percent of the female Army officers are second lieutenants (the lowest grade) whereas only 11 percent of the male officers are at that rank. Only 1 percent of the female officers are colonels, compared to 5 percent of the male officers.

Members of Congress were pleased when the Defense Department broadened the career opportunities for women but cautioned against reverse discrimination. One Con-

Coast Guard woman in a survival suit prepares a mooring line onboard a search and rescue/law enforcement boat (Official US Coast Guard photo).

gressman cited the Marine Security Guard as an example. Marine women are not sent to countries where they would not be accepted in their professional role because the local women do not work outside the home. This means that female embassy guards are not sent to some Middle Eastern countries but can be stationed in such glamorous cities as London and Paris. Therefore, male marines get a disproportionate number of the less desirable posts.

The problem also surfaces in the Navy, where there are increasing numbers of sea assignments because of the buildup in the fleet but not a comparable buildup of positions ashore. It is becoming more and more difficult to make sure that men in the Navy are not spending enormous percentages of their time at sea.

The source of combat exclusion laws was the Women's Armed Services Integration Act of 1948, legislation that authorized women to choose a military career. The Defense Department and leading military officers strongly opposed the restrictions, but several Congressmen refused to pass the measure unless specific exclusions were written into it. As the years passed, however, both sides changed positions, and as the services closed more doors to women the Congress found itself prodding them to open up.

Contemporary battle doctrine bears little resemblance to that of World War II, thus making combat exclusion laws outmoded and irrelevant as originally conceived. Few can define a combat-related position, and there is no clear-cut way to assess risk. Future battles will not be fought by hordes of infantrymen in face-to-face engagements. There are likely to be attacks in rear areas to destroy logistical bases. That is where our women will be. Is it combat or not? Women cannot fly fighter planes, but they fly the tankers that refuel the fighters. It's more effective for the enemy to shoot down the tankers. Is flying tankers a combat job or not? Women are assigned to missile silos miles to the rear of the front lines. They will be far from the "combat zone," but they will be primary targets.

It appears that most people have already accepted the fact that women will be casualties in spite of the often heard argument that public opinion is opposed to women in dangerous positions. In fact, evidence indicates that the American people are quite comfortable with the idea of women in nontraditional military roles. A 1986 poll conducted by "NBC News" concluded that 52 percent of the public supported women in combat support roles, and 77 percent were comfortable with the policy that military women will not be evacuated during a military conflict.

A similar poll taken in 1982 determined that:

- 84 percent supported increasing or maintaining the number of women in the armed services.
- 93.7 percent approved of women nurses serving in combat zones.
- 83.4 percent approved of women as military truck mechanics.
- 72.7 percent approved of women as jet transport pilots.
- 62.4 percent supported women as jet fighter pilots.
- 59.2 percent supported women as missile gunners.
- 57.4 percent approved of women as crew members on combat ships.
- 35 percent approved of women in hand-to-hand combat.

Recent history backs up the results of these public opinion surveys. We heard no great outcry when it was learned that Air Force women flew planes that delivered

Chaplain's assistant Senior Airman Jody Laws prepares the base chapel altar for a Sunday service (Official US Navy photo).

equipment and supplies to US forces during the Grenada invasion in 1983 and that Army women served in a variety of roles from the second day of the invasion. There was no outrage when in 1986 Navy women pilots flew delivery flights to carriers at sea and female Air Force pilots supported the jet airplanes that attacked Libya.

In 1987, when the USS *Acadia* entered the Persian Gulf to repair the USS *Stark*, which had been damaged by a missile attack, the American public hardly noticed that 25 percent of the crew were women sailors. Women

astronauts have died; Army women were aboard the C-141
that crashed in Newfoundland in 1986; a Navy woman was
killed in a terrorist bombing in Italy in 1988; and two
female marines were killed in a helicopter crash in 1989.
The argument that the "the American people" will not
tolerate it can no longer be used to keep women from
nontraditional, dangerous, combat-related career fields.

The question of physical strength usually enters into the
equation. Generally, women are smaller, weaker, and have
less endurance than men. Nevertheless, some women are
stronger and more aggressive than some men. It can be
argued that assignments should be based on physical
standards rather than on gender.

Women are very conscious of the perception that they
cannot hold up their end of the load. A female Navy pilot
assigned to fly transport planes in the Antarctic was told
she wasn't strong enough. She felt a lot of pressure to do
well. When her male colleagues said, "This is a man's
world," she was convinced that she had better become a
closet weightlifter. As the date for the mission approached,
her commander warned her that she wouldn't be strong
enough to handle the arctic environment, and he offered to
take her to the gym to show her how to get into shape. On
the first try, she lifted her body weight.

When a female marine sergeant reported for training as a
firefighter to rescue pilots from burning aircraft, a gunnery
sergeant complained that she wasn't strong enough to lift
his desk. She picked up one end of his desk and dumped
his papers on the floor. "Gunny," she said, "never tell a
woman she can't do something."

And in January 1989, the Coast Guard's first female
rescue swimmer, Aviation Survivalman Third Kelly M.
Mogk, pulled an entangled pilot from the sea. She was 5-
feet-6 and 115 pounds and he was 6-feet-5, 215 pounds. For

twenty-five grueling minutes she fought a strong wind, three-foot breaking seas, and sixteen-foot swells while making numerous dives to cut rigging and shrouds away from the injured Air National Guard pilot. It was her first rescue mission.

Each branch of service has its own unique mission, equipment, weaponry, and personnel needs, and for no logical reason the Navy, Marine Corps, and Air Force have more legal constraints than the Army and Coast Guard. All of these factors give a slightly different slant to how each service integrates its female members.

One of the most interesting and perhaps least publicized examples can be found in the United States Coast Guard. The Coast Guard, smallest of the armed forces, is an agency of the Department of Transportation during peacetime. Its missions include:

- Port safety
- Boating safety
- Defense preparedness
- Search and rescue
- Aids to navigation (lighthouses)
- Merchant marine safety
- Environmental protection
- Maritime law enforcement.

In time of war the Coast Guard becomes part of the Navy and has served with distinction in every war. Coast Guard units may also be assigned to Naval Task Forces as the need arises, as was the case during the Vietnam conflict.

Because it is a small service with so many important tasks, the Coast Guard counts on each member to perform to maximum potential. Most people are expected to be

Air Force T-38 instructor pilot stands in front of her aircraft (Official US Air Force photo).

multimission-capable. Women serve in assignments in all Coast Guard mission areas.

No restrictions exist on training, assignments, or career opportunities for women in the Coast Guard. The combat restriction laws do not apply. Women are assigned to all units that can provide reasonable privacy in berthing and personal hygiene. That includes shore units at home and overseas and ships that sail to both the Arctic and Antarctic and across the Atlantic and Pacific Oceans.

The women who are assigned to flight training undergo that training in an unrestricted status and upon completion are assigned to any operational command in an unrestricted flying status.

The Coast Guard Commandants have testified before Congress that they believe that the women shipboard crew members are an integral part of the crew and that their removal would be detrimental to ship operations. Therefore, in wartime, unless the Secretary of the Navy decides differently, Coast Guard women may perform in jobs from which Navy women are excluded.

The issue of whether the Coast Guard will remove women from its ships during war is debated every time the armed forces conduct a major exercise. An angered former Commandant of the Coast Guard, Admiral James S. Gracey, is quoted as saying, "I flatly refuse to [remove them]. If I have a woman commanding a ship in the Bering Sea, I am not about to bring that ship home, nor am I about to fly a helicopter out there to pick up its commanding officer just because she happens to wear a skirt once in a while. Forget it."

The Air Force has already had to deal with a problem similar to the one addressed by Admiral Gracey. The Air Force deployed women as flight crew members during the 1983 invasion of Grenada. An estimated two dozen women helped airlift troops and supplies into Grenada while US forces were engaged in combat with Cuban forces at the Point Salines Airport.

No one was injured and no aircraft was hit during the airlift operation to and from the airport. When questioned, military leaders told the newspapers that rapid planning did not give officials time to pick aircraft crews. Female pilots, flight engineers, and loadmasters were used as an expedient.

"To have excluded an aircraft from the mission simply because there was a woman aboard would have lessened our response and reduced our effectiveness," said Major General William Mall, who commanded the first wave of air forces to hit the island on October 25.

It is likely that *expediency* will have a lot to do with how women are employed in future conflicts, regardless of any written, understood, implied, or other restrictions.

The Navy probably has the most difficult problem in providing facilities for mixed crews. That was one of the points brought out by the commanding officer of a Fleet Air Reconnaissance Squadron in an April 1989 article in *Naval Proceedings*. The members of her squadron fly planes whose mission is to provide a survivable communications link to US strategic nuclear strike forces. She admitted that it is often an emotional issue:

" . . . heads [rest rooms] that previously had been allocated for only male officers or chief petty officers had to be designated male and female. This perceived loss of perks [privileges] can generate many complaints. We also shared heads by placing 'flip' signs on the door—one side read 'men' and the other 'women'.

"Our barracks had rooms with two to four beds and a private bath, with each room exiting onto a common

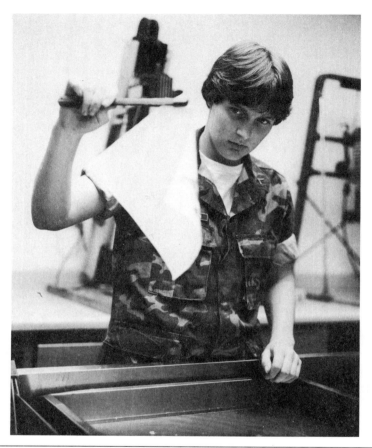

Marine Private First Class Lynn Holmgren, a photographer at the training and audiovisual support center at Parris Island, inspects a photograph before its being washed (Official US Marine Corps photo).

corridor. All occupants within a room had to be of the same sex, but rooms were mixed by sex along the same hall. Our people were mature, and these arrangements caused no particular problems, although we did maintain a regular rigorous barracks inspection program.

"Our EC-130 aircraft originally were designed in

the 1950s, long before there were female aircrew members, and the head facilities were primitive, even for the men. But our women quickly devised innovative and inexpensive ways to convert these facilities to convenient (if equally primitive) use for women."

The commander concluded that, "Facilities can be provided anywhere, even on board small ships, given a combination of money for conversion, leadership to solve turf battles, and a 'can-do,' mature attitude within the organization."

The Marine Corps has the highest ratio of combat-to-support duties, with much support coming from other services. Roughly 60 percent of its jobs are classified as combat support, compared to about 3 percent in the Air Force. This accounts for the smaller number of women in the Corps.

The Marine Corps is the rapid deployment branch of our military force. It must be ready to move on short notice, usually in the banned-for-women combat ships, presenting special problems in their assignment. When setting quotas on the number of women it can have in enlisted job specialties, it must balance combat requirements against equal promotion opportunities for everyone.

For every Marine Expeditionary Force of about 48,000 marines, the Corps must be able to field two of its three brigades with only men, who can go to sea aboard combat amphibious ships. The combat exclusion law is at work again. The Marines figure that its third brigade could be airlifted, and the combat exclusion rule would not apply.

Despite its reputation as a macho band of brothers, women with seven to ten years in the Marine Corps say that attitudes have improved since they enlisted, and younger women say they are largely satisfied.

At one time the Army conducted research studies during combat exercises to determine the level at which the addition of women to a unit resulted in decreased combat readiness. Instead, the results pointed out that women performed well, that the key to a unit's performance was not the male/female ratio, but the quality of leadership.

The Army is the largest and the most heterogeneous of the services. It is responsible for most of the land warfare, but it has its own ships and airplanes. The Army does not have the same legal restrictions as the Navy, Marine Corps, and Air Force; it can determine the assignment policies for its soldiers, and it relies heavily on a probability of risk code. Currently, about 90 percent of the Army's skill categories are open to women, but because the heavily populated specialties such as infantry are off limits, women may fill about 52 percent of the jobs.

For years rumors persisted that in the event of a military crisis in Europe the female soldiers would be evacuated. This rankled both the men and women involved. The Secretary of Defense has announced, very clearly and strongly, that this will not be the case. The fact is that Army women in Germany have been trained to use the FM-16 rifle, throw hand grenades, dig foxholes, and use antitank weapons.

The services haven't been eager to broaden the opportunities for women, but because of the dwindling pool of eligible quality male recruits and the elimination of the draft, the armed forces by necessity must meet personnel demands with increasing numbers of women. The combat exclusion laws are not likely to be changed soon, but according to Linda Grant DePauw, editor of *MINERVA: Quarterly Report on Women and the Military*, peacetime application is more theoretical than real. She observes: "The new beefed-up training of Marines in North Carolina is as 'down and dirty' for the women as it is for the men."

CHAPTER ◇ 9

Sexual Harassment:

It's No Joke

Historians find evidence of sexual harassment as far back as the 18th century. Until recently, however, this type of misconduct was seldom reported. Embarrassment, ignorance, intimidation, degradation, and fear kept victims from talking. Often the fears were justified. The most common reaction to reports of sexual harassment was nothing—the next most common response was to take action against *her*!

Regrettably, the armed forces of the United States cannot claim enlightened leadership in solving the problem. For decades sexual harassment of women in the services was not taken seriously. One after another, the Secretaries of Defense issued statements that sexual harassment would not be tolerated. Yet it was tolerated, and to a lesser extent it is thriving in some commands today.

Verbal abuse, obscene gestures, dirty jokes, pornographic posters, offers to trade advancement opportunities for sexual favors were often dismissed with excuses such as,

"Boys will be boys," "It was just a joke," or "Who asked her to join this man's army?"

It is not surprising that the first glimmers of change came in the turbulent 1970s, when feminism hit its stride and the end of the draft forced senior military leaders to view the role of women very differently. At last, women began to speak up in voices that could be heard. They did not like the jokes, the gestures, the innuendos, and they refused to smile sweetly while being insulted. Their reaction baffled many men who had never considered this type of behavior offensive.

In the 1970s the military awoke to the fact that this was far more than a social issue. Sexual harassment makes women unhappy and uncomfortable. Unhappy women who are uncomfortable at work leave and look for a job somewhere else. Therefore, sexual harassment is costly because it results in the loss of experienced and trained people. It is very expensive to recruit and train more women to take their place, and after a few years the cycle of loss starts again. Like so many other social issues, the impetus for change came from economics, not conscience.

In August 1980, Secretary of the Navy Edward Hidalgo issued explicit guidance to the Navy and Marine Corps, stating that sexual harassment was "unacceptable behavior." He charged the leaders with responsibility for ensuring that any instance of sexual harassment be dealt with "swiftly, fairly, and effectively."

Secretary of the Army John O. Marsh, Jr. sent the following memorandum not long afterward:

SUBJECT: Department of the Army Policy on Sexual Harassment

Each of us in the Department of the Army has a responsibility for maintaining high standards of honesty, integrity, impartiality, and conduct to assure the proper performance of the Army's mission.

Sexual harassment violates those standards, undermines interpersonal relationships, and interferes with the effectiveness of the force.

Sexual harassment is defined as (1) influencing, offering to influence, or threatening the career, pay, or job of another person—woman or man—in exchange for sexual favors; or (2) deliberate or repeated offensive comments, gestures, or physical contact of a sexual nature in a work- or duty-related environment.

Sexual harassment is unacceptable behavior. It is a violation of the high standards of conduct that I expect from all personnel at every rank and grade engaged in the mission of the Department.

Individuals who are sexually harassed by supervisors, superiors, coworkers, or peers should make it clear that such behavior is offensive and report the harassment to the appropriate supervisory level. It is the responsibility of every supervisor and manager— military and civilian—to examine the matter and take necessary action to insure that instances of sexual harassment are addressed swiftly, fairly, and effectively.

Complaints of sexual harassment may be filed with the Military Equal Opportunity Office, Inspector General, EEO Counselor, or Civilian Personnel Officer, as appropriate.

I know that you will support my continuing commitment to equal opportunity by exhibiting the highest professional behavior and courtesy that the nation expects from the Army.

At a higher level, Secretary of Defense Caspar W. Weinberger issued increasingly stronger messages to the Secretaries of the Military Departments on the subject. Compare the tone of his 1981 memo to the one written in 1986.

1981: "I would appreciate it if each of you would issue a strong statement promulgating DoD and service policy on sexual harassment. Your statement should inform all personnel of the avenues to seek redress and the actions that will be taken against persons violating the policy. The ASD (MRA&L) [Assistant Secretary of Defense] will review the results of your policy and report the results to me annually."

1986: "In some instances the Chain of Command does not appear either to address adequately these issues nor to respond in appropriate fashion to complaints of sexual harassment and discrimination.

Moreover, the problems of sexual harassment and discrimination continue to persist . . . We must be more diligent in our efforts to deal with these issues, understanding that sexual harassment and discrimination have potentially devastating implications for unit morale, job performance, force retention, and ultimately the readiness of the US Armed Forces.

While I am encouraged by the efforts of each of the Services to deal with these issues, we must do more to eliminate sexual harassment and discrimination within the Armed Forces."

In the end, it was the DACOWITS that put sexual harassment of military women on center stage for all to examine. After a fact-finding tour of Navy and Marine Corps bases in the Pacific area in August 1987, the DACOWITS delivered a damning report stating that servicewomen were subjected to "abusive behavior" ranging from verbal abuse to professional neglect.

Most of the Navy and Marine women who spoke to the group said that the Navy base at Subic Bay in the Philippines catered only to men. Prostitutes were free to move around the base and enter the military clubs, making the situation intolerable for servicewomen who had nowhere to go for recreation. The city outside the base had 5,000 registered prostitutes, and as a result Monday morning talk on the job revolved around the weekend sexual exploits of the men. According to the report, most women said they learned to either ignore these stories or to tell their male colleagues that they did not want to hear them.

Women complained of being grabbed, while off base, by military men who treated them as though they were "fair game."

Meredith Neizer, special assistant to Defense Secretary Weinberger, who accompanied the DACOWITS on the trip, asserted, "The quality of life for women at these bases is difficult . . . The leadership of the bases have accepted this environment and leave it to the women to deal with."

While much of the report was shocking, perhaps the most notorious incident was the allegation of "bizarre and aberrant" behavior toward female crew members of the salvage ship *Safeguard*, including a complaint that her commanding officer once jokingly broadcast via radio an offer to sell the women aboard.

There were eighteen women among the *Safeguard* crew of ninety. The women complained to DACOWITS of sexual harassment, unfair treatment, verbal abuse during crew meetings, and suggestions of sexual favors that would "make life easier for them."

The women tried to resolve their grievances through the ship's chain of command without success but had not made

their intolerable conditions known off ship until the
DACOWITS group arrived on board. Response was swift.
In September, Rear Admiral Robert T. Reimann, Com-
mander of the Middle Pacific Surface Group, relieved the
Safeguard's commanding officer. Admiral Reimann said,
"Preliminary [investigation] results substantiate sufficient
misconduct to cause me to have lost confidence in [Lt.
Commander] Harvey's ability to command."

The accusation that Harvey tried to sell the female crew
members to Korean sailors was determined to be an
"insensitive and tasteless joke," but the former commander
was punished for sexual harassment, conduct unbecoming
an officer, dereliction of duty, and fraternization. He received a punitive letter of reprimand, was fined $3,133.80,
and was reassigned to staff duty ashore.

The ship's first lieutenant and deck division officer was
issued a punitive letter of reprimand, and the ship's command master chief and master diver was reprimanded for
immoral behavior for committing an indecent act with a
female civilian in the presence of male and female crew
members.

The initial allegations were made on August 6, and the
case was closed by early October.

But the storm surrounding the DACOWITS tour was
just beginning to gather strength. The Pentagon's top
manpower official, Dr. David J. Armor, announced the
formation of a senior-level task force to assure that military
commanders overseas adhere to policies on the proper
treatment of women. He termed the findings of the
DACOWITS group "morally repugnant," and said, "It's
emphatically the Defense Department's policy that this
kind of sexual harassment will not be allowed."

At the same time, Secretary of the Navy James H. Webb
ordered Admiral Carlisle A. H. Trost, Chief of Naval

Operations, and General Alfred M. Gray, Commandant of the Marine Corps, to convene separate study groups to solve the problems of women in the sea services.

Pressure to improve conditions for women came from both sides of Congress. Less than two months after the DACOWITS trip, the House Armed Services Subcommittee on Military Personnel and Compensation launched a series of hearings that the chairman, Representative Beverly Byron, said were aimed at getting the services to improve conditions voluntarily for women. The Congresswoman threatened that if changes were not made she was prepared to introduce legislation that would force the services to act.

WHAT IS SEXUAL HARASSMENT?

How can you tell the difference between harmless male-female behavior such as flirting and sexual harassment?

Noticing that some people are male and some are female; flirting or joking; telling someone they look nice; and even asking someone out (if you are prepared to take no for an answer) are normal and acceptable conduct. They are not examples of sexual harassment unless one person forces unwelcome attentions on another.

Sexual harassment can be categorized as either the "quid pro quo type," which means trading one favor for another, or the "hostile-environment type."

Quid pro quo sexual harassment occurs when a senior offers to change a subordinate's job conditions based on sexual favors. Under the Department of Defense definition, if a person who rejects such offers from a superior is deprived of career advancement, pay, or promotion or is subjected to unusually harsh job assignments, the superior is guilty of sexual harassment.

Hostile-environment sexual harassment is often more difficult to identify. It can include verbal abuse, physical contact, or indecent actions and gestures. This type of conduct creates an intimidating, uncomfortable, offensive atmosphere and can interfere with a person's work performance.

In plain language, the following behavior is considered inappropriate:

- Using profanity
- Telling off-color jokes
- Making sexual comments
- Pinching
- Deliberate bumping or brushing against someone in passing
- Grabbing
- Touching
- Kissing
- Mauling
- Hugging
- Cornering
- Patting
- Squeezing
- Questions about one's personal activities
- Wolf whistles
- Suggestive compliments on physical attributes
- Use of endearments such as "honey" or "dear"
- Leaving sexual notes
- Using obscene gestures
- Displaying sexist cartoons and pictures

All of the military services have come out often and strongly against sexual harassment, but it is a fact of life in the armed forces. Perhaps it is inevitable because of the imbalance in the numbers. Maybe it is because so many service members live and work together so closely that their private and professional lives get tangled. Surely there is an element of machismo that causes some men to resent women who invade this last male preserve. Whatever the cause, you would have a hard time finding a servicewoman who has not either experienced sexual harassment or known someone who did.

The question to ask yourself is, "Is this career worth it?" If the answer is yes, then you have to know how to handle the problem and how to cope.

There is no Defense Department-wide policy on resolving sexual harassment complaints. Sexual harassment is classed with discrimination against race, religion, and national origin, and it is left to each service to make plans to combat unlawful discrimination.

In 1987 the *Navy Times* published an article, "Your Guide to Sexual Harassment," which gave the following advice to servicewomen:

- If a woman believes she is being harassed by a peer because of her gender, she should confront that person. Sometimes saying "Knock it off, buddy" works.
- Telling someone to curb his actions is not so easy, however, when rank is involved. If a man of superior rank sexually harasses a woman, she should take her complaint to his immediate superior. If the result is unsatisfactory to her, she should continue taking her complaint up the chain of command—to the base or post commander if necessary.
- At that point, the woman can file a nonformal "referral" grievance. This generally requires filling out a form to be submitted to the commanding officer,

asking him or her to look into the complaint and decide whether or not a formal investigation is warranted.

- If the commander decides that the woman's complaint is invalid and chooses not to investigate, the woman may file a formal complaint if she is still not satisfied. If the formal complaint is against someone other than the commanding officer, the woman submits the complaint to the commanding officer. If the complaint is against the commanding officer, the woman submits the complaint to the offending man's commanding officer.

All members of the armed forces are subject to a set of rules called the Uniform Code of Military Justice. Those who break the rules, such as someone accused of sexual harassment, are tried by a military court called a court-martial. This is very serious, and a person found guilty has much to lose.

The chart below gives examples of violations of the Uniform Code of Military Justice related to sexual harassment:

If the harasser	*He may be guilty of*
1. Threatens to influence adversely the career, salary, or job of another in exchange for sexual favors	Extortion, assault, communicating a threat
2. Offers rewards for sexual favors	Bribery and graft
3. Makes sexual comment and/or gestures	Indecent, insulting, or obscene language prejudicial to good order, provoking speech, or gestures, disrespect

Anyone found guilty of these offenses can face punishments ranging from informal counseling to fines to dismissal from the service, depending on the seriousness of the charge.

The system is in place and the advice is good, but there are flaws. Many women are rightly afraid to submit a complaint about their immediate supervisor. They fear that they will get into more trouble. Others believe, like a black Air Force officer, that when you start making accusations about racism, sexism, or harassment everyone scatters, and formal complaints should be a last resort.

Most females seem hesitant to use the system. Officers are less likely to take advantage of it, but they are in a better position to deal with the problems.

A former Naval attorney offers some coping suggestions to women who pursue formal complaint procedures:

Seek out contacts in the civilian community such as Battered Women Centers, Rape Crisis Centers, the NOW (National Organization for Women) local chapter.

Go through the military system, but do it with the support of a civilian resource, whether it be an attorney, a social worker, or a religious group. Trying to do it alone will be overwhelming.

If the complaint is not redressed in a reasonable amount of time, contact your congressional representative.

Keep a diary of the offending incidents. Document the date, time, place, what was said, who was there, and what happened.

If you are required to give a statement to an investigating party, get an attorney, preferably a female.

Expect to have your records examined very closely.

WHAT DO THE WOMEN SAY?

Servicewomen who have spoken out in articles and books have been very candid on the subject. They seem to agree that confidence, competence, and a sense of humor are the weapons of choice. It helps to have grown up with brothers and to have participated in team sports, but neither is essential. Trying to be one of the boys or taking on masculine characteristics is almost always fatal. It's just as bad to play the role of the over dependent sweet young thing. Teasing and excessive flirting are unfair, dangerous, and unprofessional because there are so few women. Forget the tears!

A woman in the New York Army National Guard reported that she was approached by telephone and solicited for "special favors" in return for promotion considerations. She immediately went to her supervisors and explained the situation, and they in turn had her contact the Equal Opportunity Office (EO). She got help, and the problem was resolved to her satisfaction.

A female Command Sergeant Major who spent twenty-five years in the Army said that there has been no decrease in sexual harassment and sexual discrimination in the Army—it has just become more subtle. She blames leaders who fail to stop or are incapable of stopping the kind of behavior that fosters sexual harassment.

This woman, who was one of only four women at the time to hold such a high grade, said there is a double standard in dealing with the problem. In her view, a junior enlisted man would be in trouble, but not an officer or senior enlisted. "Nothing is going to happen to them unless it becomes an embarrassment to the Army."

In December 1984 a young Navy dental technician reported two instances of sexual harassment committed by a Rear Admiral. She charged that he made suggestive remarks to her while she was cleaning his teeth and on another occasion became aggressive. She reported this to her supervisors. The woman was given a lie detector test and transferred to another Naval base. During the next few weeks the Admiral was put on leave, and the accusations were investigated. He subsequently returned to his duty station, packed his bags, and submitted a request for early retirement, which was approved before the end of January. On February 1 he received a written reprimand.

A female sailor wrote to the *Navy Times*:

"Enlisted female personnel who take their claims of harassment to their senior enlisted advisor are told that they [the women] are probably the cause of the 'problem' and not to 'rock the boat.' Female service-members' and civil service employees' claims of harassment taken to the equal opportunity officer are often ignored. Claims taken up the chain of command all too frequently result in no visible action taken against the offender.

"Rather the norm is retribution directed against the woman or women making the claim."

In *Sound Off!* by Dorothy and Carl J. Schneider, one servicewoman said, "I just determined that you had to develop selective hearing and somewhat of a tough skin."

Another said, "Anything they talk about, I can talk about. I used to read a lot, so I could fit in there. I wanted to be able to talk about football, or basketball. After they found out that I wasn't going to cry if they called me names, that I was going to get up and call them names back, they started leaving me alone. I really feel I have more respect than some of the male NCOs in the unit."

A marine who started out to act like a man learned quickly that that wouldn't work because " . . . they'd laugh at you behind your back—or in front of your face. As time went on I found that trying to act like who I am and what I look like, I'd have much more success."

Training is the key strategy used by all the services to combat sexual harassment. Formal classes are given at all levels to teach men and women how to recognize and deal with it. Most consider it a leadership issue that must be dealt with from the top. If the senior officers let it be known that they will not tolerate such behavior, and if reported cases are handled properly, there is seldom a problem.

Many of the women who have been successful learned how to deal with men. For some there was a trial and error period. They found that they must prove their competence and must separate their professional life from their personal life. At work they must carry their own weight.

Brigadier General Gail M. Reals, the second woman in the Marine Corps to achieve that rank, believes that basic training built her confidence. She learned how to endure the treatment—the catcalls and rude remarks—many women marines get from some of their colleagues. Although she says that she never experienced sexual harassment from the men she worked for, she does remember the experience of walking down the street near a large Marine base.

"I always said to myself, 'They must be talking about someone else . . . that's not me personally, so to hell with them,'" she said. "I was not going to be driven out of the Marine Corps by people who had no stake in my life."

Army Brigadier General Myrna Hennrich was interviewed by the New York *Times*: "Some military women say that most sexual harassment today is verbal; has that been your experience?" Her answer—"No, it has not been my experience; but then I'm told I have a witty and sharp sense of humor, and in many cases I might beat people to the punchline."

The authors of *Sound Off!* concluded that, "Common sense and some experts suggest that as men and women get used to working with each other not only in the same work environments but in the same jobs, they develop new ways of interacting as colleagues. The military world encourages this by dressing women and men similarly or identically, by housing women and men in the same barracks or bachelor officers' quarters, by insisting that females are soldiers before they are women, by putting more women to work in traditionally male jobs, by paying women the same as men of their rank, by placing women in positions where they supervise men, and by formally instructing their members about what constitutes sexual harassment and discrimination and punishing it. As a result, the services have probably moved further toward the reduction of overt and open sexual harassment than most civilian employers."

Captain Delia Waldrop, United States Army Reserve, wrote:

A MESSAGE FOR FEMALE SOLDIERS ONLY

We are women who chose to be soldiers; we were not expected to deny our femininity when we put on the uniform. However, femininity should never be flaunted, abused, or used to gain special treatment.

Professional conduct begets professional respect. When we conduct ourselves with dignity, self-respect, and respect for our fellow soldiers—male and female— we foster an atmosphere of dignity and respect that will influence those around us.

While the Army clearly defines to us what jobs we can hold; establishes our standards of appearance; and

evaluates our performance, potential and professionalism, our duty descriptions do not contain the words
"fragile," "helpless," or "sex object." We are equals
among equals and should expect to be treated as such.
We are soldiers who possess and contribute significant
talents and abilities which make the Army Reserve the
vital part of the Army's total fighting force that it is
today.

CHAPTER ◇ 10

Marriage, Mates, and Motherhood

Military family policies were sexist for decades. Right into the sixties there was a lot of grumbling but a relatively quiet acceptance of what many people believed was a "woman's role." Agreement was nearly universal that a woman's responsibilities as wife and mother took precedence over her career, and that if the two came into conflict the career could be sacrificed.

The Department of Defense has come a long way in adjusting its views to the realities of modern life-styles. Pentagon officials have directed the services to develop administrative procedures that address marriage, pregnancy, children, and single parenthood. It is no longer a strictly female problem. Thousands of single male parents in uniform have custody of their children, and with so many wives working, the military can no longer assume that every married soldier has a woman at home to take over when he is called away on short notice.

At the highest levels the rules are reasonable and women have little to complain about. But, as so often happens, it is at the lowest level—on the job—that a woman may come up against ill feelings about family matters. How successfully she copes depends on the attitude of the commanding officer, the first line supervisor, her coworkers, and very often how she handles herself.

Family life in the military carries its own unique pressures that civilians never experience. Even under the best conditions—ideal regulations, an understanding boss, supportive coworkers, solid marriage—the military wife/mother must be prepared to face overseas assignments to remote places with poor health care available for herself or her children; unexpected alerts that take her away for hours or months with little advance notice; shift work; extended hours; temporary duty away from home base; orders to locations where husbands and children are not permitted; field exercises; sea duty; and an acute shortage of child-care centers.

Marriage and motherhood should never be taken lightly, but there is reason to be even more thoughtful about making these important decisions if one or both partners are in the service. The needs of the Army, Navy, Marine Corps, Air Force, and Coast Guard always come first. That is the way it has to be if the country is to have an effective and strong defense. There will be no time to find a new baby-sitter when a midnight call to report to the ship is received. Mothers cannot bring sick toddlers to work.

People do manage, and many do it beautifully. About one third of servicewomen are married, and most are married to servicemen. Some cope easily, others find that something has to give. Some decide not to have a family. Some eventually split up. If one gives up the career, it is usually, but not always, the woman. Statistics show, however, that marriage, pregnancy, and divorce rates for military members are not far different than for civilians.

In an unusual turnaround, military minds now recognize that by supporting family needs they can improve readiness. In 1988 the Navy and Marine Corps sponsored a three-day conference to examine the typical problems faced by service couples. Compare this to the attitude that persisted into the seventies: " . . . if we wanted you to have a family, we would have issued you one." The Assistant Commandant of the Marine Corps told the conference, "Mission readiness is not limited to equipment repair and supply availability. A marine with his or her mind on family concerns is not giving 100 percent to the job at hand."

We will look at the history of family policies in the armed forces before discussing the issues of marriage, pregnancy, and children.

EVOLUTION OF MILITARY FAMILY POLICY

Until the manpower shortages caused by the war in Vietnam, women, but not men, were allowed to leave the service because they were married. That caused an unacceptable loss of trained females and resentment among the men. It was sexism with the men on the receiving end. The Marines, fully committed to combat in the sixties, were the first to tighten the rules; married women could no longer leave voluntarily unless they were stationed too far from their husband to establish a joint household. Even then, they had to have been separated at least eighteen months. All the services soon followed with similar rules.

This marked the beginning of the establishment of joint assignment procedures. For the first time the services made a real effort to place husbands and wives at or near the same base. It may sound like common sense today, but it was a radical departure in the sixties.

Dependency.   The next policy to be challenged was the blatantly discriminatory definition of dependency. By custom, the armed forces prided themselves on care of family members, known for many years as dependents. Servicemen with dependents received extra money and were assigned to apartments or homes on the base. The dependents, generally wives and children, were issued identity cards that permitted them to enter the base, receive medical care, shop, and use recreational facilities.

A woman was not allowed to claim dependents unless she could prove that her husband or children were in fact dependent on her for more than half of their support. For a husband to be considered a dependent, he almost always had to be mentally or physically handicapped. Being a student or retired did not count, because the thinking then was that an American family was supported by the male. At the same time, a serviceman could be married to a millionaire who would automatically be afforded all the benefits of a dependent. In the view of the Director of Women Marines in 1972, this inequity was the primary complaint among women in the Corps.

It took a decision of the US Supreme Court to settle the problem. An Air Force lieutenant married to a civilian student claimed that the policy amounted to unreasonable sex discrimination. In 1973 the justices agreed and said that the services would either have to provide the same benefits to women or revise the rules to require men to prove dependency of their wives and children. The Department of Defense subsequently declared that women would be treated equally with men in all matters of dependency and that in the future the word "spouse" was to replace "wife" and "husband."

Pregnancy. In 1948 a military study group wrote:

> "It is believed that a woman who is pregnant should not be a member of the armed forces and should devote herself to the responsibilities which she has assumed, remaining with her husband and child as a family unit."

For two decades the services followed this line of reasoning and separated all women who became parents either by birth or adoption; had personal custody of a child under eighteen years of age; or had a stepchild in the home more than thirty days a year. Thus, rules about pregnancy and children were the next to be tested.

The first break came in the late sixties when the services began to grant waivers to selected women with adopted children or stepchildren. Still, not many expected that women with their own children would be allowed to remain in service and continue with their career. The Director of Women Marines astounded those attending the 1970 Women Marine Association convention by suggesting the idea of natural mothers on active duty.

It took a few determined women and the federal courts to make her prediction come true in 1976. The debates were heated. Men and women worried about the impact on defense. One old soldier was heard to complain, "It's no way to run an army."

Many feared that women would lose a great deal of time away from their duties for pregnancy and child care. In 1977 a study of time lost by service members concluded that the differences for men and women were not significant because statistically men lost much more duty time than women for unauthorized absences, alcohol/drug abuse, auto accidents, hospitalization resulting from fights, and confinement to the brig.

At the time no one was ready to suggest the need for maternity uniforms. When a woman grew too large for her uniform, she reported to duty in civilian clothing.

MARRIAGE AND MATES

Officials claim that the number of military members wed to each other has grown to over 110,000. Much of the increase is attributed to the rising numbers of women entering the services in a wider range of career fields.

It has caused major poblems for military planners, who must meet mission goals and at the same time try to accommodate the couples who want to be stationed together. The higher the rank and the more specialized the career fields of the husband and wife, the more complicated the problems become.

The Army has developed an automated Married Couples Program, in which both soldiers in a marriage enroll. About nine months before time for one of them to be reassigned, the computer generates both names so that administrators can try to send both to the same installation or to ones within fifty miles of each other.

The objective is to keep couples together, but the first responsibility is to the commander in the field. Although about 80 percent of the couples are in the same relative geographic area, the number of separated couples is expected to increase because of the increase in dual career marriages.

There is also reasonable concern about the potential conflicts of interest and appearances of favoritism when couples are assigned to the same unit. All services forbid one spouse to directly or indirectly supervise a mate, but at times the lines in the chain of command are blurred.

Although only about 7 percent of all male military members have military wives, more than one third of all married military women are part of a dual-military-career family. Almost all women end up marrying military or staying single. Not many civilian men want to follow a wife around the world for twenty years.

In these mixed marriages, former military husbands adapt more easily than nonveteran husbands. The ideal situation in a military-civilian union seems to be when the man is a retired serviceman who has completed his twenty or more years of active duty. He contributes to family support with his pension; he is entitled to all benefits and has free access to the base in his own right. Further, he may enjoy helping and encouraging his mate as she progresses through her career.

A civilian husband without military experience runs the risk of feeling like an outsider. He has no base privileges except through his wife; his identity card carries her name; the name over the door on military-provided housing is hers. He may find it difficult to join in at social gatherings while his military wife makes easy conversation with the other men. Unless he is fortunate enough to have a profession such as writer or free-lance photographer, he will be looking for work every few years and will not be able to pursue a long-term career.

On the other hand, it is easy to see the advantages of a dual-military-career marriage. The partners truly understand the demands of each other's job and can exchange shoptalk. They take great pride in their accomplishments.

Coast Guard maternity uniform (Official US Coast Guard photo).

The wife is less likely to feel like an outsider in social situations, and when they move to a new location they have nearly identical support systems to make the adjustment with ease. It usually takes a civilian spouse longer to fit into the new community than the service member who goes to work and is immediately part of a close organization. Perhaps the most significant benefit enjoyed by military women married to military men is the ability to establish a career. Unlike the civilian wife, she does not have to change jobs every few years and is not bound to accept whatever little position happens to be available near her husband's duty station.

Some couples purposely join the military together because of the career opportunities, the chance to travel together, and the monetary benefits. Each spouse is entitled to a housing and food allowance, so it usually happens that they take home more money than if married to a civilian.

A disadvantage that should not be overlooked is that of career competition. This issue plagues career-oriented couples both in and out of the service. No matter how modern the marriage, there is potential trouble when a woman climbs the ladder of success faster than her husband. If the woman is more ambitious than her spouse, she has to think about the consequences of applying for advanced schooling that will enhance her opportunities. It is not unusual for an enlisted woman to turn down the chance to become an officer because of the likely strain it would bring to an enlisted/enlisted marriage.

All things considered, some researchers feel that women in a military/military marriage are among the happiest in the armed forces. The key is to make wise decisions about marriage, know what the challenges are, and learn to be flexible.

MOTHERHOOD

None of the social issues facing the armed forces has created more emotion and controversy than the pregnancy of active duty soldiers, sailors, marines, airmen, and coast guardsmen. In today's Army, pregnant women can remain in service and are expected to return to duty six weeks after giving birth. Many military supervisors worry that pregnancy dangerously threatens mission readiness and morale.

The chairwoman of the DACOWITS acknowledges that the pregnancy issue is a serious one, but she contends that male perceptions of female lost time due to childbirth are exaggerated. A Navy study begun in 1988 seems to confirm her view. Only 10 percent of Navy supervisors said that motivation among women sailors declines when they become pregnant. More important, on the average pregnant sailors lose only about one day a month for prenatal care or pregnancy-related illness. Nevertheless, these same supervisors also said that having them at sea hurts military readiness.

Managing a military unit with pregnant troops is a challenge not eagerly accepted by many. The pregnant women are often limited in what they can do; they have regular medical appointments; perhaps they cannot stand for long periods of time; eventually they will not be able to lift heavy objects; their coworkers may have to bear part of their load; and they will probably be away from work for five or six weeks.

The supervisor has to accommodate these needs and be fair to the rest of his people. An Army captain said that in one year his unit had practically every possible combination with men, women, and babies—wanted, unwanted, planned, simple and complicated, overdue and premature. He learned that each pregnancy is different and that the way it affects work depends largely on the woman and how the people around her are able to deal with it. If the soldier was hard-working and dedicated before her pregnancy, she won't change. If she always looked for ways to get out of work before, she has a great escape now—if her leaders allow it.

The situation is sometimes better handled by women supervisors who have themselves been pregnant. They know that pregnancy is not an illness or a disease and should not be used to shirk responsibility. One woman officer said that before she had children she pampered her pregnant subordinates. Now they are less mothered.

The attitudes of managers often reflect the type of work done by the unit. It is far easier to accommodate a pregnant marine who works in an office than one who routinely has to climb telephone poles. An extreme example was reported by a veterinarian who told of the problems encountered when one of his "good soldiers" became pregnant. The doctor said that she could not lift any animal weighing more than 20 pounds, assist during radiographic procedures or surgery using gas anesthesia, handle cats, nor check cat feces for parasites. She had to stay away from formaldehyde, a potential cancer-causing agent.

In the Navy women can stay with their ship until the 20th week of pregnancy or until the ship goes to sea. But they cannot get under way with their ship unless they can be medivaced within three hours to a facility that handles childbirth. After giving birth, they must return to the ship after four months and complete their tour of duty.

In the Air Force women pilots are not allowed to fly. In the Marine Corps commanders may, with the approval of a doctor, shorten convalescence leave after childbirth in cases where the marine's absence from duty has a clear adverse impact on the operational mission of her unit.

Some people who fill out work schedules agonize over the problem of convalescence leave, feeling that four to six weeks is too generous. Many hold that the extra time off for a new mother is unfair because of the burden it places on those who have to take over her duties. Others, especially the pregnant women, think four weeks is too short. In fact, the armed forces policy is to permit up to six weeks of leave after the baby arrives.

It is estimated that at any one time between 7 and 9 percent of all military women are pregnant. That is comparable to civilian statistics for women of the same age group. The threat to mission readiness clearly depends on the organization. Units that are overseas and closest to danger are most affected; the Navy has unique problems because of the demands of shipboard duty.

Maternity uniforms are a common sight at any military installation today. Pregnant women are given a supplemental uniform allowance to purchase them. In 1988 maternity fatigues, complete with the familiar brown and green blotches, were introduced into the supply system. They look a lot like the uniforms worn in combat, except that they have a front stretch panel and the top blouses over the midsection. The women think they're great, but some veterans say the Defense Department has finally gone too far. In making the announcement about the camouflage uniforms, a Marine Corps spokesperson said, "Many women may be assigned to duties even in advanced pregnancy which are not excessively strenuous but do require uniforms which provide comfort and ease of maintenance."

In perspective, it is recognized that women are an integral part of the all-volunteer force. To discharge them for pregnancy would create a severe manpower shortage and would deplete the armed forces of highly trained and experienced troops who would have to be replaced with less capable recruits. For now, at least, it seems that the military must face the issue and continue to try to manage its operations with the presence of pregnant troops.

The dilemma does not end when the baby is born and the mother returns to duty. The shortage of adequate child-care centers in the military is considered a major concern. A member of Congress recognized it as a readiness issue. In an emergency, the armed forces must move quickly, and in the mother's absence who will take care of their children?

The majority of married servicemen can count on their civilian wives to handle the responsibility. That is not available to the services' growing number of single or dual-career parents.

The problem surfaces in periods of peace as well as war because:

- Slightly more than half of all US military installations do not have organized child-care centers.
- The US bases that do offer on-base child-care centers meet only about 60 percent of the demand.
- One third of all children whose parents request spaces in child-care centers are on waiting lists.
- Less than 20 percent of the centers are open at night when military people are required to work.
- Military women are allowed six weeks of maternity leave after giving birth, but one third of all day-care centers at US military bases do not accept infants under six months.
- The cost of child care is out of reach of many junior enlisted parents, especially if they are single.

Both house of Congress sponsored legislation aimed to ease these problems, not because of their social implications, but because they adversely affect the nation's ability to effectively wage war or defend itself against attack. We can look forward to more money being appropriated for construction of day-care centers. More and better-trained child-care workers will be available to manage the centers to meet the needs of military parents.

The authors of *Sound Off!* have this to say:

"While their situation falls short of the ideal, American military women fare better than most of their civilian sisters in maternity leaves, guaranteed jobs after childbirth, and child-care facilities. The servicewoman may wish she had more time to recuperate and to bond with her new baby, but she can count on four weeks [now six] off and a job to which to return, without loss of rank, seniority, or pay. Her doctor can lengthen her disability leave. And if her job and her boss permit, she may save up her leave beforehand so that she can extend her time at home after her baby is born . . . The wide world of the military makes room for all sorts of families, and many kinds of women. But no more than the civilian world has it resolved the ultimate problem of the pull between family and career."

CHAPTER 11

The Service Academies

I n the fall of 1976 women enrolled in the famed service academies at West Point, Annapolis, Colorado Springs, and New London. The Congress of the United States made it happen.

A furor erupted when in 1972 Senator Jacob Javits of New York nominated a woman to the Naval Academy at Annapolis, only to learn that she was not even considered. He did not take the rejection quietly, and his public statements fired the imagination of his colleagues on Capitol Hill. Meanwhile, the Navy decided to integrate the Naval Reserve Officers' Training Course (NROTC), in a move some thought was designed to deflect attention from the idea of women at Annapolis.

Interestingly, the senior women service directors were split on the question, with the Navy and Army opposed and the Air Force and Marine Corps in favor. Navy Director Captain Robin Quigley explained:

"The [Naval] Academy exists for one viable reason, to train seagoing naval officers and also to give the Marine Corps a hard core of career Regular officers. There is no room, no need, for a woman to be trained in this mode, since by law and by sociological practicalities, we would not have women in those seagoing or warfare specialties."

General Mildred Bailey of the Army agreed, adding:

"We don't need to send women to the Academy to get sufficient qualified women into our officers' program. We get all we need at no expense to the government. Why should we spend money to train them?"

General Jeanne Holm of the Air Force expressed a different view:

"In my personal opinion, I would like to see women in the Academy in the not too distant future."

Lawsuits brought by women who wanted to enter the service academies and public hearings in Washington kept the debate lively. In 1974 all the service chiefs, taking the cue from Deputy Secretary of Defense William P. Clements, Jr., were vocally opposed. An official Defense Department statement was published:

"The primary responsibility of the Department of Defense is to provide for the National Defense. The Service Academies, in providing officers to fill combat roles in the Armed Forces, are essential ingredients of that national defense. Training cadets at the Academies is expensive, and it is imperative that these opportunities be reserved for those with potential for combat roles."

Air Force Academy Superintendent Lieutenant General Albert P. Clark embellished the proposition that the sole focus of the Academies was to prepare leaders for combat:

"The environment of the Air Force Academy is designed around these stark realities [of combat]. The cadet's day is filled with constant sports, rugged field training, use of weapons, flying and parachuting, strict discipline and demands to perform to the limit of endurance mentally, physically, and emotionally. It is this type of training that brings victory in battle. It is my considered judgment that the introduction of female cadets will inevitably erode this vital atmosphere."

The reality is that the academies have always been regarded as something more than combat leader training institutions. They provide an excellent education and entree into the elite of the armed forces. Graduates are promoted faster and get choice assignments and more command opportunities. Representative Samuel S. Stratton revealed that of the 8,880 graduates of the Air Force Academy on active duty in October 1974, 29 percent had never had a career combat assignment. In fact, at the time more than 10 percent of the graduates of the Army, Navy, and Air Force Academies had never had a combat assignment.

No one mentioned the practice of giving deferments to some Academy graduates, the athletes who delayed their service obligation to play professional football. The Air Force had the weakest position to defend, since it accepted

Cadet Kristin Baker is the first woman to be selected as brigade commander for the US Corps of Cadets, United States Military Academy, West Point (US Army photo by Spec. Mike Weber).

applicants who were not qualified to fly and could therefore never qualify as Air Force combat leaders.

Proponents were equally vocal. Representative Donald Fraser, who nominated a woman sailplane pilot to the Air Force Academy and had her application returned unconsidered, vehemently exclaimed:

> "How cynical that while we make an effort to recruit women into the forces—to quadruple their number by 1977—we are denying them access to the best educational program of their profession. We are squandering our human resources."

Representative Fortney H. (Pete) Stark, Jr., who also had a woman's nomination returned, testified:

> "In short, women should, with haste and without question, receive appointments to the academies. The fact is that we need them there—with their talents and skills and perspective—far more than they need the academies."

Representative Patricia Schroeder observed:

> "Women should go to the academies for the same reasons men go—to pursue a military career, to be pilots, to get a good education."

During subsequent congressional hearings, as pressure was increased by the legislators, the services agreed with varying degrees of enthusiasm that if Congress so ordered, the academies could be integrated. It was rumored at the time that the Superintendent of West Point said something like " . . . over my dead body."

Unexpectedly, but characteristically, in 1975 the Commandant of the Coast Guard announced that in July of the following year the Coast Guard Academy at New London, Connecticut, would accept women, ending a hundred-year tradition.

The signing of Public Law 94-106 on October 7, 1975, ended the debate for the three larger academies. It did not surprise anyone that the Army sent rather dismal brochures to young women who requested information whereas the Air Force sent letters to every high school in the nation encouraging bright women to apply. Consequently, at the end of the first summer of training the highest level of success was at the Air Force Academy; the lowest level was at West Point.

But West Point was soon in step and in the summer of 1989, based on academic excellence, athletic abilities, and military skill demonstrated during her first three years at the Point, Cadet Kristin Baker was named First Captain. As the top cadet, commander of West Point's brigade of 4,400, she follows in the footsteps of men like Douglas MacArthur, John "Black Jack" Pershing, and William Westmoreland.

Baker and her female colleagues have taken their places in one of the last holdouts of male dominance in the military. At the time of her historic appointment, women made up about 10 percent of the armed forces and 10 percent of the cadet classes at the Army, Navy, and Air Force Academies.

WHAT TO EXPECT

All service academies offer four years of college education leading to a bachelor of science degree. The majors available at each institution reflect the unique mission of the branch of service; maritime and oceanographic subjects at the Naval Academy and aeronautical subjects at the Air Force Academy. Majors are offered in such areas as political science, mathematics, and the humanities. Women cadets have had a history of choosing academic majors in a far different pattern than the men, electing the humanities and social sciences over the technical courses. Some fear that this trend will limit them as they progress through their careers.

A significant difference between an education at a service academy and a civilian university is that academic excellence by itself in not good enough to achieve success. Leadership, military performance, and physical performance are integral to the program, and their importance cannot be underestimated. It is understood by all that cadets must demonstrate the highest ethical and moral standards.

Generally, the women's leadership and military performance grades have lagged behind the men in the plebe (first) year, but they tend to catch up and often surpass the men. This may be the consequence of their limited preparation in high school, where women may have had fewer competitive experiences. Although the women catch up in areas of military development by graduation, in their post-academy careers the female graduates are heavily represented in support and administrative assignments. This is more likely the result of their selection of a course major than a reflection on their leadership ability.

In their first year women show a surprising aptitude for physical education classes, often rating higher than their male classmates. Knowing that the physical demands are going to be tough, they seem to prepare themselves. By graduation, however, more of the marginal women are apt to drop lower while more marginal men improve.

Naval Academy plebes grimace in the heat of tug o'war competition during plebe summer. First Class midshipmen (at right) shout encouragement. Participation in sports and athletics is key to success for women at the service academies (US Naval Academy photo).

The physical environment is uncompromising and arduous. Women who are not physically fit do not survive. Female athletes not only fare measurably better but are better accepted by the men. The athletes are more physically fit and exhibit highly desirable traits such as teamwork, self-discipline, confidence, and endurance. At the Naval Academy the record shows that only 12.7 percent of the women who have won a varsity athletic letter have failed to graduate. In comparison almost 50 percent of the women who never participated in an Academy varsity sport left before graduation.

Studies have identified a number of problems, disappointments, and obstacles peculiar to women students at the academies. Male students show varying degrees of acceptance of the women, and often their opinions and negative attitudes are rooted in the fact that because women are not allowed to hold combatant jobs, their contribution to the service is limited.

There are few role models for the women, as few staff members are women who have had command assignments. As more academy graduates advance in their own careers and return as faculty, it is hoped that this situation will be eased.

The women are really a minority, and most have been adamant that they want to be treated precisely like everyone else. Yet some feel that a woman's support group would be beneficial, as the female students have no one with whom to discuss the unique problems they face. There are indications that instances of sexual harassment, discrimination, and prejudicial attitudes have occurred and gone unnoticed by the authorities.

It is not unusual for male students, especially in the first year, to engage in inappropriate behavior such as trying to make the women cry, using derogatory language, telling offensive jokes, or singing demeaning marching songs.

The academies continually evaluate the process of integrating women into the system, and they want to do it right. The goal is to graduate well-prepared military officers, the next generation of leaders. Special attention is being paid to the reasons female cadets and midshipmen drop out, fail, or leave the service after completing their five-year obligation.

Gaining entrance into the academies is not easy, but the real challenge begins when you get there. The rewards, professional and personal, are unparalleled.

Finances: Cadets (Army, Air Force, Coast Guard) and midshipmen (Navy) receive tuition, medical care, room and board, and approximately $500 a month that pays for uniforms, books, and incidental expenses. Upon graduation, the newly commissioned officers are obligated to serve on active duty for at least five years.

Eligibility: Admission to any of the academies requires that the applicant be:

- at least 17 years of age (but not have passed the 22nd birthday by July 1 of the year of entrance);
- a citizen of the United States;
- of good moral character;
- able to meet the academic, physical, and medical requirements;
- single;
- free of obligation to support dependents, including parents.

Only the finest of the nation's student leaders are selected. It is never too soon to get ready for an academy education. Here are some things you can do while in high school.

Academic Preparation: the following courses are recommended:

- four years of English;
- four years of college-prep math;
- four years of social studies;
- four years of lab science;
- three years of social studies;
- two years of foreign language;
- one year of typing.

Special consideration is given students with honors and those studying advance courses. You can improve your qualifications with educational experiences such as gifted and talented programs, exchange student programs, and summer computer and science programs.

Good study skills are essential. The key factors in academic evaluation are your high school grades, class standing, and SAT or ACT scores (with emphasis on math).

Leadership Preparation: In leadership evaluation, the academies consider participation in student government and clubs, community involvement, scouting, and after-school employment.

The person who excels in a few select activities generally ranks higher than the person who participates in many activities but excels in none. The military training portion of an academy education focuses primarily on leadership, making any experience in this area valuable.

Physical Preparation: Your physical fitness will be measured with a special candidate fitness test, a thorough medical examination, and a vision test. You must have no history of illegal drug use. Medical examinations are scheduled following evaluation of academic qualifications.

Your record of high school athletic achievements is

considered. To prepare for the tough physical demands you will face at the academy, you should do regular distance running, develop your upper-body strength, and be able to swim. Strength and endurance are necessary for your success.

The following information regarding the nomination process is extracted from the Department of Defense publication *PROFILE*, dated January 1989.

NOMINATIONS

Applicants for the Army, Navy, and Air Force Academies must have a nomination in order to be considered for appointment. It is best to apply for all nominations for which a person is eligible. However, no nomination or appointment is needed for the Coast Guard Academy. Selections are made by the Academy directly, based on scholastic aptitude and nationwide competition. Prospective applicants should request a precandidate questionnaire from the service academy at the time they apply for their nomination. This will initiate an admission file.

Most applicants receive nominations from members of the US Congress; however, other nominations are available through affiliations with the armed forces. The types of nominations applicants may apply for include:

Congressional: These nominations are available to anyone who meets the basic eligibility requirements. Applicants may apply to the Representative and Senators from their legal home of record. It is not necessary to know the congressional member personally. Each congressional member and the Vice President may have five constituents in attendance at each of the academies at any one time and may nominate as many as ten applicants for each vacancy.

The spring of the high school junior year is the suggested time to apply for congressional nominations, though some members of Congress will accept late requests for nominations in November and early December of the senior year.

Presidential: This nomination is available to children of career members of the military services, regular or reserve, if the parent is on active duty and has served continuously for at least eight years; is retired from active duty; or is a deceased retired veteran. By law, a person eligible under the Children of Deceased or Disabled Veterans category (see below) may not be a candidate under the presidential category.

Regular and Reserve Unit: Nominations are allowed for enlisted men and women in regular and reserve units of the Air Force, Army, Marine Corps, and Navy. Enlisted personnel may apply only for the service academy of their respective military component in this nominating category, although they may apply to the other service academies in other categories. People serving in any service may apply for the Coast Guard Academy.

Children of Deceased or Disabled Veterans: This category is for children of deceased or 100 percent disabled veterans whose death or disability was determined to be service-connected, and for children of military personnel or federally employed civilians who are in a missing or captured status. Applicants holding a nomination under this category are not eligible for nomination under the presidential category.

Honor Military/Naval School: This category includes distinguished graduates from these sources.

Administrator, Panama Canal Commission: Residency is required.

Delegates to Congress from American Samoa, Guam, and the Virgin Islands: Residency is required.

Delegates to Congress from the District of Columbia: Residency is required.

Governor and Resident Commissioner of Puerto Rico: Residency is required.

ROTC: Students from junior and senior ROTC units may request nominations through their school detachments.

Note: A separate appointment category is open to children of Medal of Honor recipients. They are appointed to the academy of their choice without regard to vacancies, provided they meet minimum qualifying standards.

Appointments to the Coast Guard Academy are made competitively on a nationwide basis. A nomination is not required. Qualified men and women may apply by writing to:

Director of Admissions
U.S. Coast Guard Academy
New London, CT 06320-4195

For more information write to:

Director of Admissions
U.S. Military Academy
West Point, NY 10996

Director of Admissions
U.S. Naval Academy
Annapolis, MD 21402

Director of Admissions
U.S. Air Force Academy
Colorado Springs, CO 80840

CHAPTER 12

It's Your Decision: Enlistment Programs and Training

"The military has been truly a rewarding experience that's going to be the basis of whatever I do in the future. No way around that. They have trained me. They have given me an opportunity to meet new people and learn from them. On the other hand, there's nothing that you can do to stop things like discrimination. You can try to control it, of course, but that's a reality. Any woman coming in today, I would tell her that 'You will in your career be discriminated against. But don't let that be the thing that determines your future. It's just like realizing that a stop sign exists at a corner. Okay. You know it's there. It might detain you a few minutes. But if you're patient, and you're willing to work around it, you will succeed in your course, wherever you're going. You just have to stop for a sign every now and then.'"

Army Reserve and National Guard Officer candidates train together in the Empire State Military Academy at Camp Smith, Peekskill, New York (NYARNG photo by Capt. Paul A. Fanning).

This practical advice was offered in an interview with Dorothy and Carl Schneider for their book *Sound Off!— American Military Women Speak Out*. It sums up the situation pretty well. Sexism exists in the military, but probably no more than anywhere else. By themselves, sexist attitudes and policies should not be the factors that keep you from considering a military career. In fact, many women would say that the armed forces offer far greater opportunities with far fewer inequities than comparable civilian professions.

Selecting an occupation that will lead to a satisfying career takes careful planning and informed decision-making. A multitude of factors must be considered: qualifications, education requirements, training offered, challenges, pay, benefits, quality of life, opportunities for advancement, travel, adventure, and yes, of course, barriers such as sexist attitudes and policies. Be sure to put it all into proper perspective. Every important decision in life involves some trade-offs.

Spend time gathering information and think about your own goals and preferences. Go to your school counselor for information and advice. By all means finish high school. Employers value people who have demonstrated the ability to persevere. Your diploma is evidence that you finish what you start.

Research has shown that high school graduates are more likely to be successful in the military than nongraduates. Therefore, the services accept very few nongraduates. That is even more true of women than men. In recent years the military had had a pool of highly qualified women applicants to select from; they are going to take the best.

The figures tell the story. On active duty today, approximately 75 percent of the enlisted women have high school diplomas, 20 percent have completed some college, and 5 percent have college degrees.

Your first challenge is to finish school. But how much schooling is necessary? Do you want to be enlisted or an officer? The two are very different in many ways, including qualification requirements, training provided, type of jobs, level of responsibility, obligated years of service, and pay. To become a commissioned officer almost always demands a college degree.

ENLISTMENT PROGRAMS

The First Steps. The *Military Career Guide*, published by the Department of Defense, outlines a four-step process to enlistment: Talk to a recruiter, qualify for enlistment, meet with a service classifier, and enlist.

Talk to a recruiter. Better yet, talk to a recruiter from every service. Even if you think you are not interested in a particular branch of the armed forces, you can learn a lot by listening and asking questions. Each one will highlight the advantages of that service. After each interview you will be better prepared to ask the right questions of the next recruiter.

Ask about training opportunities, career fields open to women, career fields closed to women, service life, enlistment options. If you are interested in a nontraditional field, ask how many women are in that occupation. Are any positions in that skill field closed to females?

The recruiter will determine your eligibility for enlistment, specific training, and assignments. Enlistment programs vary greatly among the five services, and they are regularly adjusted to meet recruiting needs. The general qualifications for military enlistment are shown in the accompanying table, taken from the *Military Career Guide 1988–1989*. Keep in mind that the specific requirements may vary from one service to another and are subject to change.

Be prepared when you visit the recruiting office. Bring a birth certificate and Social Security card or other proof of citizenship and date of birth. Applicants born overseas of American parentage are required to provide proof of citizenship. Aliens must provide proof of lawful entry for permanent residence.

The second step, qualifying for enlistment, usually takes

General Enlistment Qualifications

Age	Must be between 17 and 35 years. Consent of parent or legal guardian required if 17.
Citizenship Status	Must be either (1) U.S. citizen; or (2) an immigrant alien legally admitted to the U.S. for permanent residence and possessing immigration and naturalization documents.
Physical Condition	Must meet minimum physical standards listed below to enlist. Some military occupations have additional physical standards.

Height—   For males: Maximum-6'8"
                                Minimum-5'0"

      —   For females: Maximum-6'8"
                                 Minimum-4'10"

Weight—There are minimum and maximum weights, according to age and height, for males and females.

Vision—There are minimum vision standards.

Overall Health—Must be in good health and pass a medical exam. Certain diseases or conditions may exclude persons from enlistment, such as diabetes, severe allergies, epilepsy, alcoholism, and drug addiction.

Education	High school graduation is desired by all services and is a requirement under most enlisted options.
Aptitude	Must make the minimum entry score on the Armed Service Vocational Aptitude Battery (ASVAB). Minimum entry scores vary by service and occupation.
Moral Character	Must meet standards designed to screen out persons likely to become disciplinary problems. Standards cover court convictions, juvenile delinquency, arrests, and drug use.
Marital Status and Dependents	May be either single or married; however, single persons with one or more minor dependents are not eligible for enlistment into military service.
Waivers	On a case-by-case basis, exceptions (waivers) are granted by individual services for some of the above qualification requirements.

place at one of the 68 Military Entrance Processing Stations (MEPS) located around the country. At the MEPS you will typically take the Armed Services Vocational Aptitude Battery (ASVAB) if you haven't already done so. ASVAB results are used to determine whether you qualify for entry into a service and have the specific aptitude level required to enter job specialty training programs. High school counselors can give you information about taking these tests locally.

While at the MEPS you receive a medical examination, which takes about two hours. Women are examined (to include a pelvic examination) separately and in privacy, with a female escort present. Height, weight, and physical standards vary by service and by sex.

The third step is to meet with a career information specialist who will help you select a military occupational field. Your ASVAB scores are used to assess your vocational potential and your academic abilities. Pay attention at this point. Always remember that the recruiter must meet the personnel needs of the service and is required to fill a quota by career field. Don't let anyone talk you into an occupation that does not interest you. On the other hand, go into the interview with an open mind. You will be presented with an array of options that you never thought about. Find out which fields offer the best chance for training and advancement. Which fields are best if you are considering a long-time military career? Which fields provide training that is easily transferable to civilian jobs? Will you receive a bonus if you enter a hard-to-fill occupation?

After talking to the classifier, you select an occupation and schedule an enlistment date. Enlistment dates may be scheduled for up to one year in the future to coincide with job training openings. This option is called the Delayed Entry Program (DEP).

Following selection of an occupation and a training program, you sign an enlistment contract and take the oath of enlistment. If you choose the DEP option, you return home until your enlistment date.

Service Contract and Obligation. Your contract specifies the enlistment program, enlistment date, term of enlistment, and other options agreed upon such as training, cash bonuses, etc. If the service cannot meet its part of the agreement (for example, to provide a specific type of training), you are no longer bound by the contract.

In return for the promised job, training, pay, and benefits, you agree to serve for a certain period of time, Depending upon the program, enlistees serve between two and six years on active duty and a period of time in the reserves that will bring the total service to eight years. The balance between active duty time and reserve time is generally determined by the amount of training required to prepare you for your chosen field.

Benefits. Military pay and benefits are set by Congress and are the same for members of all services regardless of sex.

The major part of an enlisted member's paycheck is basic pay, which is determined by pay grade (rank) and length of service. Special additional pay is generally awarded to individuals serving under special or unusual conditions, such as flight, sea, or hazardous duty.

Most enlisted members, at least for the first few years, live in military housing and eat in military dining halls. Those who must live off-base in civilian quarters are paid a tax-free allowance to cover housing and food costs. Additional allowances may be paid to individuals who live in high-cost areas of the country or world.

A clothing allowance, which is different for men and women, is paid to enlisted members for upkeep and replacement of military clothing. Travel and transportation allowances are paid to all service members when assigned to a new duty post.

Enlistment bonuses of up to $8,000 may be awarded to qualified people enlisting for certain skills for a specified length of time. Reenlistment bonuses may be paid to enlisted members who complete their enlistment and choose to reenlist.

A summary of employment benefits is contained in the accompanying table, published by the Department of Defense in *Military Career Guide 1988–1989*. These benefits change from time to time, so it is always a good idea to check with the local recruiter.

Basic Training. Basic training, demanding and rigorous, is designed to turn civilians into disciplined soldiers, sailors, marines, airmen, and coast guardsmen in a very few weeks. It is not easy, and you will be bone-tired. But if you are in reasonably good shape, did well in school, can follow instructions, and are motivated, you can make it. A very high percentage of young men and women do.

Recruits are formed into groups of 35 to 80 for training. These groups are called companies in the Navy and Coast Guard, flights in the Air Force, and platoons in the Army and Marines. Experienced enlisted persons conduct the training and instruction. The training groups tend to become closely knit teams and develop group pride and camaraderie.

The non-commissioned officer, in some services called the drill instructor (DI), who is responsible for your training will push you to your outer limits. DIs are always tough, and they invariably shout a lot. It will only seem that you are their only target; nearly everyone will be thinking the same thing. As one Army brochure promises, "You'll come

Summary of Employment Benefits for Enlisted Members

Vacation	Leave time of 30 days per year.
Medical, Dental, and Eye Care	Full medical, hospitalization, dental, and eye care services for enlistees and most health care costs for family members.
Continuing Education	Voluntary educational programs for undergraduate and graduate degrees or for single courses, including tuition assistance for programs at colleges and universities.
Recreational Programs	Programs include athletics, entertainment, and hobbies:
	Softball, basketball, football, swimming, tennis, golf, weight training, and other sports
	Parties, dances, and entertainment
	Club facilities, snack bars, game rooms, movie theaters, and lounges
	Active hobby and craft clubs, book and music libraries.
Privileges	Exchange and Food, goods, and services are available at military Commissary stores, generally at lower costs.
Legal Assistance	Many free legal services are available to assist with personal matters.

to remember your Sergeant [DI], if not with affection, certainly with respect."

Both men and women can expect indoor and outdoor classes covering the following subjects:

- Military Customs and Courtesies
- Physical Fitness
- Confidence Course
- Drill and Ceremonies
- Wearing the Uniform

National Guard Sergeant Tracy Zoch (seated) and Specialist Andrea Burrall process personnel records. Like their active duty counterparts, guardsmen are trained in the use of modern hi-tech equipment (NYARNG photo by Capt. Paul A. Fanning).

- Inspections
- Guard Duty
- Code of Conduct, Geneva and Hague Conventions
- Marches
- Familiarization with US Weapons
- Service Benefits
- Military Justice
- Equal Opportunity
- Hazards of Drug and Alcohol Abuse
- Personal Health and Hygiene
- Nuclear, Biological, and Chemical Defense
- First Aid

- Field Hygiene and Sanitation
- Defensive Training
- History and Traditions

Recruits assigned to combat arms receive additional training in offensive tactical training.

Physical fitness and stamina are developed and maintained through daily exercises, competitive sports, and marches. Pushups, sit-ups, jumping jacks, distance running, water survival, and swimming instruction are all part of the routine. Tests are given to measure the degree of physical fitness attained. Physical fitness standards differ for men and women. Most services recognize a difference in upper-body strength between the sexes.

A recruit's day begins early—usually at 5:30 a.m.—and ends at 9:30 p.m. There is little free time, and travel away from the training group is limited, if permitted at all. Saturdays and Sundays are sometimes lighter; trainees are always given time to attend church services. Recruits may receive visitors at certain times.

Basic training is usually conducted in two sections: recruit training and job skill training.

US ARMY: Men and women in the Army are called soldiers. Army basic training is given in a number of locations, including training centers in New Jersey, South Carolina, Georgia, Texas, Kentucky, Alabama, Oklahoma, and Missouri. Men and women receive essentially the same initial training, including weapons instruction, but may be trained separately.

US NAVY: Enlisted men and women in the Navy are called sailors. Navy recruit centers are located in Orlando, Florida; Great Lakes, Illinois; and San Diego, California. Women recruits train only at Orlando.

In addition to the general military subjects common to

all services, Navy instruction includes courses in aircraft and ship familiarization and basic deck seamanship.

US MARINE CORPS: Men and women in the Marine Corps are called marines. Men undergo training, usually referred to as boot camp, at Parris Island, South Carolina, or in San Diego, California. All women train at Parris Island.

Although both males and females receive intense periods of instruction in defensive combat, only males receive instruction in offensive combat. A man's career begins with eleven weeks of recruit training followed by four weeks of advanced field and combat skills at the School of Infantry at either Camp Lejeune, North Carolina, or Camp Pendleton, California. Women receive a combined package of recruit training and follow-on field skills training while at Parris Island over a twelve-week period. Schools for the more technical specialties follow the stint at Parris Island.

US AIR FORCE: Enlisted men and women in the Air Force are called airmen. All men and women receive the same basic training during six weeks at Lackland Air Force Base, Texas. After basic, most airmen attend a resident course at one of the Air Force's five technical training schools.

US COAST GUARD: Enlisted men and women in the Coast Guard are called coast guardsmen. All Coast Guard recruits attend recruit training (boot camp) at Cape May, New Jersey, for approximately eight weeks. In addition to courses common to all services, coast guardsmen study seamanship, ordnance, and damage control.

After completion of boot camp, coast guardsmen are assigned to their first unit and are encouraged to apply for a skill school. Most complete at least six months at the unit before being sent to a school. Those who do not request a school can advance through on-the-job training.

OFFICER PROGRAMS

Commissioned officers are the leaders of the military. They are the managers, executives, and professionals (doctors, nurses, lawyers, etc.). Officers set the goals, develop the plans, and lead junior officers and enlisted in putting the plans into action. Officers are responsible for the well-being, morale, training, and quality of life of all those who serve in the same unit. Their responsibilities go far beyond the duties of their occupational field.

To join the service as a commissioned officer, you must have a four-year college degree. Certain scientific, technical, and professional fields may require an advanced degree. In addition, specified mental aptitude, physical, and moral standards must be met. The *Military Career Guide* summarizes the general qualification requirements for military officers in the following table. Specific requirements may vary by service and for certain occupations. For accurate, up-to-date information, you must contact a recruiter.

General Officer Qualifications

Age	Must be between 19 and 29 years for OCS/OTS; 17 and 21 years for ROTC; 17 and 22 years for the service academies.
Citizenship Status	Must be U.S. citizen.
Physical Condition	Must meet minimum physical standards listed below. Some occupations have additional physical standards.

	Height—	For males:	Maximum-6'8"
			Minimum-5'0"
		For females:	Maximum-6'8"
			Minimum-4'10"

Weight—There are minimum and maximum weights, according to age and height, for males and females.

Vision—There are minimum vision standards.

Overall Health—Must be in good health and pass a medical exam. Certain diseases or conditions may exclude persons from enlistment, such as diabetes, severe allergies, epilepsy, alcoholism, and drug addiction.

Education   Must have a four-year college degree from an accredited institution. Some occupations require advanced degrees or four-year degrees in a particular field.

Aptitude   Must achieve the minimum entry score on an officer qualification test. Each service uses its own officer qualification test.

Moral Character   Must meet standards designed to screen out persons unlikely to become successful officers. Standards cover court convictions, juvenile delinquency, arrests, and drug use.

Marital Status and Dependents   May be either single or married for ROTC, OCS/OTS, and direct appointment pathways. Must be single to enter and graduate from service academies. Single persons with one or more minor dependents are not eligible for officer commissioning.

Waivers   On a case-by-case basis, exceptions (waivers) are granted by individual services for some of the above qualification requirements.

Commissioning Programs. Commissioned officers are obtained from five sources:

- Service Academies
- Officer Candidate School (OCS) and Officer Training School (OTS)
- Reserve Officers' Training Corps (ROTC)

- Medical Programs
- Direct appointment

Service Academies. Information on the four service academies can be found in Chapter 11.

Officer Candidate/Training School. College graduates can become commissioned officers in the five services without prior ROTC or military training through OCS or OTS. There are several versions of this type of program. See a recruiter in your freshman year of college to get the latest facts. The standard Officer Candidate/Training School begins after graduation.

Depending upon the service, training may take up to twenty weeks. After successful completion, you are commissioned as a military officer and have a service obligation of four years.

Reserve Officers' Training Corps. Undergraduates in nearly 1,400 public and private colleges and universities receive training to become military officers through the Reserve Officers' Training Corps. ROTC, the source of nearly a third of the military's new officers each year, did not become co-ed until 1969. It is an exceptional opportunity; depending upon the service and option selected, as an ROTC candidate you can receive scholarships for tuition, books, uniforms, and a monthly allowance.

The training consists of several hours of classroom instruction and drill each week on campus and some summer training periods at a military base.

Upon graduation you are commissioned a military officer and can choose to fulfill your obligation either on active duty or in the reserve or national guard.

Medical Programs. The armed services offer health care specialists financial assistance, scholarships, and medical degrees in return for specified periods of military service. Numerous options are available for nurses, doctors, dentists, veterinarians, dietitians, medical technologists, optometrists, pharmacists, psychologists, physical therapists, and health care administrators.

There are awards for students as well as persons who already have baccalaureate degrees. Some of the programs offer the full pay and benefits of an officer even while in school.

Medical commissioning programs are offered by the Army, Navy, and Air Force. There are no medical officers or enlisted persons in the Marine Corps or Coast Guard.

For more information see a recruiter or write to:

Assistant Secretary of Defense (Health Affairs)
The Pentagon
Washington, DC 20301

Director of Admissions
Uniformed Services University of Health Sciences
4301 Jones Bridge Road
National Naval Medical Center
Bethesda, MD 20014

Air Force ROTC
Office of Public Affairs
Maxwell Air Force Base, AL 36112

United States Air Force Health Service
HQ USAFRS/RSH
Randolph Air Force Base, TX 78150

Chief Nurse Corps USAF (SGN)
Bolling Air Force Base
Washington, DC 20332

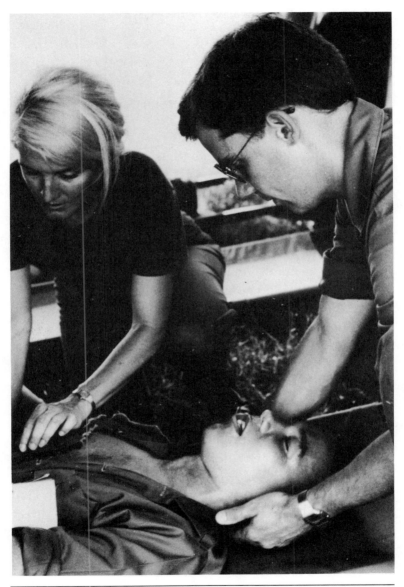

Air Force nurses stabilize a patient's neck and treat a simulated chest wound during the battlefield nursing course (Official US Air Force photo).

Chief Army Nurse Corps (DASG-CN)
5 Skyline Place
5111 Leesburg Pike
Falls Church, VA 22041

Director Navy Nurse Corps (OP-093N)
2100 Potomac Annex
23rd and E Streets NW
Washington, DC 20037

Women in the health care fields in the armed forces face many of the same problems as their sisters in civilian medical careers, especially in the less traditional roles as doctors, veterinarians, etc. There is one important difference: They generally enjoy wider acceptance as military professionals than do nonmedical servicewomen. There is, in particular, great respect for women nurses, notably among men who have been wounded in combat.

Though it sounds cynical, because female nurses are not in competition with men (except male nurses), they are viewed as less of a career threat than nonmedical servicewomen. Nurses are administered separately and compete among themselves for rank, pay, and training.

As you might guess, probably the greatest discrimination in military health care is felt by the male nurses.

Direct Appointment. Professionally qualified persons such as doctors, nurses, lawyers, chaplains, and civil engineers may take advantage of the system of direct appointments, and they may be eligible to enter at a higher rank than the usual entry-level grade of ensign (Navy and Coast Guard) or second lieutenant (Army, Marine Corps, Air Force). Educational prerequisites differ and the selections are competitive.

Basic Officer Training. Depending on the program and service, officer training may take place while in college or after graduation. You can expect strenuous physical conditioning, calisthenics, running, forced marches, field exercises conducted under primitive conditions, and difficult confidence courses. Officers are expected to be role models and be able to do whatever their enlisted troops can do. It would be foolhardy not to prepare yourself for this aspect of the training. Women in all branches of the military feel pressured to do well physically because of the perception of weakness and of not "pulling their weight."

The topics covered in training include:

- Role and Responsibilities of the Officer
- Military Law and Regulations
- Service Traditions
- Military Customs and Courtesies
- Leadership
- Career Development
- Military Science
- Administrative Procedures
- Military History

After completion of basic training, you will attend a Basic Officers Course and/or specialized training for your occupational specialty. As a commissioned officer you can expect periodic professional military training and the opportunity to earn advanced degrees.

RESERVE AND NATIONAL GUARD FORCES

Seven separate forces make up the reserves—the Army, Navy, Air Force, Marine Corps, and Coast Guard Reserves and the Army and Air National Guard. The reserves are critical to our national defense because they can be called to active duty in an emergency.

Reservists and guardsmen receive basic and skill training on active duty, then return home to attend regular drill periods, most often scheduled one weekend a month. Additionally, they attend at least one two-week active duty session per year.

Pay is based on a scale according to grade, length of service, and the number of days allotted for each training period. Members of the reserve and guard receive one day's pay for each training assembly performed. A normal weekend drill is considered as four assemblies and entitles members to four days of basic pay.

During peacetime the reserves perform many functions to support the active forces and often participate in joint training operations, even overseas. Approximately one third of our nation's defense units are in the reserves.

The generous education benefits are often cited as the reason for enlisting in the reserves. Enlistees are eligible for benefits up to $5,040 when certain conditions are met.

Reserve membership is also an option for people who have completed their military obligation and want to return to civilian life, yet continue their affiliation with the armed forces. They can advance in rank, gain valuable job skills, enjoy the camaraderie of other service members, add to their income, and earn a retirement.

Nearly 39 percent of the women in military service are in the reserve and national guard. The reserve is an alternative to full-time active duty for women with families.

For more information, see a recruiter. Often you will find the reserve and national guard recruiters at local armories.

SUMMARY

So, you see, it's all up to you. If you are interested in the military, you have some homework to do and some options to weigh. Finish high school. See your guidance counselor. Talk to as many recruiters as possible. Consider your career and education goals. Then make an informed decision. If you decide to go for it, recognize at the start that you will come up against some challenges. But, then, an interesting life is full of challenges—some of them are definitely worth taking on.

CHAPTER 13

It's About Time:

Women Veterans

An honorable discharge or retirement papers did not ensure women the same veterans' benefits as men. It is true that in the beginning only a few entitlements were specifically denied to most women; these usually had to do with dependents and survivors. Sexist and discriminatory attitudes followed women out the gate like an invisible shadow.

In the minds of too many people, including the women themselves, they were not veterans because they did not engage in hand-to-hand combat. Yet no one questioned the status of men who served as cooks, clerks, or telephone operators.

At the conclusion of World War II some female veterans were told that since they had enlisted for patriotic reasons, it would be inappropriate to expect anything in return. Others were turned away from veterans hospitals where the staff were not prepared to deal with female medical problems. Still others ran into some sticky legal barriers.

The Women's Army Auxiliary Corps (WAAC) was formed in 1942 as a civilian unit attached to the Army. At the time, one of the fears was that if women were part of the real Army, their husbands would expect a government allotment whether needed or not because civilian wives of soldiers received an allotment whether they needed it or not. The WAACs thought they were in the Army, as did the men soldiers and the American people. After all, most of the administrative procedures used to rule the WAAC were the same as those used to govern the male soldiers.

But in some situations they were considered civilian federal employees and as such were not entitled to veterans benefits. The WAAC was reorganized in 1943, and those women who transferred to the Woman's Army Corps were protected by The Department of Veterans Affairs (DVA) regulations. The women who did not make the transition had to wait for special legislation in 1980 to become eligible for their benefits.

The Women Air Force Service Pilots were an elite group of female pilots who in World War II flew nearly everything while delivering new planes, retrieving crippled aircraft, and towing targets for the men. They suffered 38 casualties out of a force of 1,047. The original intention was to militarize them if the program was successful, and since it was widely acclaimed, everyone took their military status for granted. These talented and brave women had to wait thirty years for their benefits.

Theoretically, all other women veterans were entitled to the same care as men. That is how the laws were written, but in practice the women were shamefully neglected. The most blatant breach of faith was in the area of health care. Because women comprised less than 2 percent of the armed forces from the 1940s to the '60s, their needs were simply not addressed.

Health studies did not include women, and although the DVA keeps volumes of statistics on men, women were not mentioned until the '80s. It was assumed that the women would not use the hospitals, or if they did, the hospital would cope. Rarely were there adequate bath or toilet facilities; privacy during examinations and hospitalization was nonexistent; and gynecoloists were not available.

Male patients in veterans hospitals have always enjoyed the services of barbers and received personal care items like razors and pajamas. The women also got their haircuts from barbers and received razors and men's pajamas. The pharmacies did not stock medications for illnesses considered "female," and the hospitals were not authorized to buy examining tables or equipment designed for women. Sanitary napkins were never available, not even in the hospital shop, so nurses kept a stock for distribution that they purchased with their own money.

Even after the DVA became sensitive to the needs of women, one hospital had a urinal instead of a toilet in the women's examining room, and another had a diagram of a man's body on the chart used by physicians during physical examinations.

Many women were told they could not be served, and many simply gave up because of the atmosphere. Clearly, they were not welcome.

In 1983 a damaging report by the Government Accounting Office documented the plight of women veterans, and since that time the DVA has made substantial improvements in the care of women veterans. Female health care, morale, and accommodations have become priorities. Gynecologists have been hired; expensive mammography machines used to detect breast cancer have been purchased; and newsletters and personal invitations have invited the women to come and use the hospitals.

Until 1972 married female veterans did not receive the same education benefits as married male veterans; widowers of female veterans were not entitled to the same death benefits as the widows of male veterans; and their survivors were not eligible for government loans for homes, farms, or businesses.

Today, with the exception of the health system, which still needs some improvements, all benefits are available to women with no real or artificial roadblocks set up to make them inaccessible.

MEMORIALS

Towns and cities across the country have their war memorials along with statues of men who fought in every skirmish since the Revolution. Few visual images of women portray their contribution to our nation's defense. A notable exception is "Molly Marine," the bronze woman who has stood at the intersection of Elk Place and Canal Street in downtown New Orleans since the dedication on the Marine Corps birthday in 1943. The inscription reads:

Molly Marine
November 10, 1943
FREE A MAN TO FIGHT
REDEDICATED JULY 1, 1966, IN HONOR OF
WOMEN MARINES WHO SERVE THEIR
COUNTRY IN KEEPING WITH THE HIGHEST
TRADITIONS OF THE UNITED STATES
MARINE CORPS

Vietnam Women's Memorial Project. Especially sensitive to the lack of recognition given women veterans, several nurses who had served in Vietnam established the

"Molly Marine," the first statue of a military woman in uniform in the United States, was dedicated in New Orleans in 1943 (Official US Marine Corps photo).

Vietnam Women's Memorial Project, intending to add a statue of a nurse clad in combat fatigues to the well-known Vietnam Memorial in Washington, DC. Success has come in small doses, and they anticipate that the memorial will be dedicated in 1992 or 1993, after a decade of removing legal obstacles and overcoming opposition.

Just when they thought they had it clinched in 1987, the Commission of Fine Arts in the nation's capital rejected the plan. One commission member said that the memorial was complete as it is and allowing the statue of a woman would open the door to others seeking representation for their ethnic group or military specialty. Some men said that adding a woman to the "Wall" was like painting the Statue of Liberty pink or adding the likeness of Elvis Presley to Mount Rushmore.

Vietnam veterans, the men who fought the war, with rare exceptions were strongly in favor of the women's statue, and the 1987 official rejection seemed to create a backlash of support. A member of Congress wrote, "The time has come, in the opinion of many of us, to complete the Vietnam Memorial with a sculpture that represents the women's experience in the Vietnam War."

The women did not give up, and in November 1989 both houses of Congress passed legislation authorizing construction of the memorial. Financial and procedural hurdles remain, but few doubt that the project will reach fruition.

Women in Military Service Memorial (WIMSA). It may have been the publicity and controversy surrounding the Vietnam Women's Memorial Project or even the attention given to the dedicaton of the Vietnam Memorial itself —when so many veterans talked of being forgotten—but whatever the stimulus all women veterans listened and said, "What about us?" Thus plans were made for a national memorial to honor all women who have ever served in the armed forces of the United States.

The stated purpose of the Women in Military Service Memorial Foundation is to:

- Pay tribute to the women who have been and are in our armed forces.
- Tell the story of their dedication, commitment, and sacrifice.
- Make their historic contribution a visible part of our national heritage.
- Illustrate their partnership with men in defense of the nation.
- Inspire others to emulate, follow, and surpass them.

The memorial will be built on six beautiful acres at the main gateway to Arlington National Cemetery. It is a choice spot near the grave of President John F. Kennedy. It will be far more than a statue, featuring a theater-style education center showing a movie about the historical contributions of women in the military; a room with the names and photos of registered women service members available for computer review; a reception area with commemorative books; and displays.

Visitors will be able to punch in the name of a friend or relative who served, and if she is on the computerized registry, facts about her service record will come up on a video screen. So will her picture in uniform if submitted. Those registering are literally writing servicewomen's history.

The two memorials are not competing with each other, despite the efforts of opponents of the Vietnam Women's Project, who would like to scuttle plans to put a nurse at the Vietnam Memorial and therefore support the WIMSA plan recognizing all women. Brigadier General Wilma Vaught, president of WIMSA, expressed the feelings of both groups when she said, "Surely there is room for two things for women in this city. Goodness knows there are enough statues of men on horseback to go around."

When American women who have served in the military hear of the proposed memorials, regardless of which one they support, they nearly always say the same thing, "IT'S ABOUT TIME."

Bibliography

Binkin, Martin, and Bach, Shirley. *Women and the Military*. Washington DC: The Brookings Institution, 1977.

DePauw, Linda Grant. *Seafaring Women*. Boston: Houghton Mifflin, 1982.

Hadley, Arthur Twining. *The Straw Giant: Triumph and Failure: America's Armed Forces*. New York: Random House, 1986.

Hancock, Joy Bright, Captain USN (Ret.). *Lady in the Navy: A Personal Reminiscence*. Annapolis: U.S. Naval Institute Press, 1972.

Hewitt, Linda, Captain USMCR. *Women Marines in World War I*. Washington, DC: History and Museums Division, United States Marine Corps, 1974.

Holm, Jeanne, Major General USAF (Ret.). *Women in the Military: An Unfinished Revolution*. Novato, CA: Presidio Press, 1982.

Keil, Sally Van Wagenen. *Those Wonderful Women in Their Flying Machines: The Unknown Heroines of World War II*. New York: Rawson, Wade Publishers, Inc., 1979.

MacCloskey, Monro, Brigadier General USAF (Ret.). *A Definitive Study of Your Future as a Woman in the Armed Forces*. New York: Richards Rosen Press Inc., 1979.

Marshall, Kathryn. *In the Combat Zone: An Oral History of American Women in Vietnam, 1966–1975*. Boston; Little, Brown, 1987.

Meid, Pat, Lieutenant Colonel USMCR. *Marine Corps Women's Reserve in World War II*. Washington, DC: Historical Branch, C-3 Division, Headquarters, United States Marine Corps, 1968.

Mitchell, Brian. *Weak Link: The Feminization of the American Military*. Washington, DC: Regnery Gateway Inc., 1989.

Rogan, Helen. *Mixed Company: Women in the Modern Army*. New York: G. P. Putnam's Sons, 1981.

Schneider, Dorothy and Carl J. *Sound Off!: American Military Women Speak Out*. New York: E. P. Dutton, 1988.

Slappey, Mary McGowan. *Exploring Military Service for Women*. New York: Rosen Publishing Group, 1986.

Stiehm, Judith Hicks. *Arms and the Enlisted Woman*. Philadelphia: Temple University Press, 1989.

―――. *Bring Me Men and Women: Mandated Change at the U.S. Air Force Academy*. Berkeley: University of California Press, 1981.

Stremlow, Mary V. Colonel USMCR (Ret.). *A History of Women Marines 1946–1977*. Washington, DC: History and Museums Division, United States Marine Corps, 1982.

Treadwell, Mattie. *U.S. Army in World War II: Special Studies —The Women's Army Corps*. Washington, DC: Department of the Army, 1954.

Willenz, June A. *Women Veterans: America's Forgotten Heroines*. New York: Continuum Publishing Company, 1983.

Index